A Navigator's Universe

Published for The Newberry Library by *The University of Chicago Press*

A Navigator's Universe

The LIBRO de COSMOGRAPHÍA of 1538 by Pedro de Medina

Translated and with an Introduction by URSULA LAMB

International Standard Book Number : 0-226-51715-2
Library of Congress Catalog Card Number : 70-128870

The University of Chicago Press, Chicago 60637
The University of Chicago Press, Ltd., London

Protected under the provisions of the International Copyright Union
Published 1972
Printed in Great Britain

A la memoria de un maestro español Don Alberto Jiménez Fraud

Contents

The manuscript of the *Libro de cosmographia* is the property of the Bodleian Library, Oxford, England. I am obliged to the Keeper of the Western Manuscripts, Dr. R. W. Hunt, who granted permission for its publication, and to the staff of the Bodleian for their unfailing courtesy over the span of pleasant years I spent with their treasures. I am grateful to Mr. H. C. Taylor for allowing me to consult his unique and hitherto unknown copy of the *Coloquio de cosmographia* by Pedro de Medina and to the director of the Beinecke Rare Book Library at Yale, Mr. Herman W. Liebert, for permission to use the collections in his care.

The directors of Spain's archives were hospitable and helpful beyond the call of duty. D. José de la Peña y Camara, who was then director of the Archivo General de las Indias at Seville, expedited microfilms which I requested; D. Ricardo Magdaleno, director of the Archivo General de Simancas, had searches made for me; and in the deposits of Madrid, I was given help wherever I turned. D. Julio Guillén, director of the Museo Naval in Madrid, shared with me some of his vast knowledge of naval literature, and D. Roberto Barreiro-Meiro helped me find information that I sought in the museum.

I would not have approached this task without the aid of Commander David Waters's *Art of Navigation in England in Elizabethan and Early Stuart Times*, which the author further augmented by his personal encouragement to me.

Both the translation of the *Libro de cosmographia* and the introductory text have benefited immeasurably from the help of Professor R. L. Colie, whose patience and empathy make of this critic by profession the gentlest of teachers. I join a company of scholars who know what I mean when I regard my debt to her beyond my means to express it.

There remains the happy task of acknowledging the moral support of the Society for the History of Discoveries, which took me into their ranks when I arrived somewhat homeless from years abroad. The meetings of this group, which take place once a year under the energetic leadership of its founders, have been a most gratifying experience.

The Yale Council for Area Studies extended financial aid to cover the microfilm expenses of this and a current project on the Spanish cosmographers. I owe a large debt to a succession of editors, Mr. John Parker, Mr. James M. Wells, Mr. David Woodward, Mr. Henry Y. K. Tom, and those of the University of Chicago Press, all of whom took on an arduous job with patience and kindness. The plan to print the translation on facing pages had to be abandoned in the production process, and the strictures of that format could not be entirely avoided.

The dedication of this book is to the memory of a great Spanish gentleman and teacher revered by many of his students, D. Alberto Jiménez Fraud. In his house in Oxford, news about the Bodleian *Libro* of the Maestro Pedro de Medina found its first welcome.

Acknowledgments

Part One

INTRODUCTION

Pedro de Medina, the author of the *Libro de cosmographía*, identified himself on the title page of his work as "Cosmographer". In his first dialogue he defines cosmography as the science of the stellar universe and the lands and waters upon the earth's surface. He leaves unexplained what a cosmographer was or what he did, presuming that was known among his contemporaries. Unlike men's thoughts, which are free to run ahead of or behind the tenor of their times, what men actually are or do and how they do it are bound to historical situations. To write about the *Libro* could be simply an exercise in ideas, but to write about Medina, the author and cosmographer, requires some historical reconstruction of his time and circumstances.

Medina lived in the mid-sixteenth century in the city of Seville, the hub of Spain's nautical enterprise and the gathering-point for ships from the New World. During his lifetime, geographic discoveries were furnishing new facts which required a reconstruction of ideas, just as in our time new facts in the physical and life sciences demand similar reassessment. Then as now, the prime task of thinking people was to determine and to understand the new facts and to connect them usefully to what was already known without being dominated either by the authorities of the past or by notions of a predetermined future.

In answer to this challenge the sixteenth century saw the emergence of a group of men dedicated to the reorientation of Western ideas within this new world: the cosmographers. It was their task to know and to explain how the universe was constituted. As astronomers and astrologers, as geographers and experts in matters pertaining to navigation and auxiliary sciences, the cosmographers tried to assimilate the new information into what was useful from the old.

In Spain, an office of cosmographer was for some time attached both to the Casa de Contratación de las Indias (India House or Board of Trade) at Seville and to the Royal Council of the Indies. There were chairs of cosmography at the universities, and the matters in which cosmographers were called upon to lend their expert judgment were many and varied. Although cosmography began as a comprehensive, ill-defined field embracing most of the ancient sciences and ended as merely the title for instruction in the disciplines of geography, astronomy, and navigation, there was a moment in history when cosmographers were the custodians of most of the new knowledge relating to navigation and exploration, which was Europe's open sesame to the riches of the wider world.

Exact observation fostered by the cosmographers corrected ideas about the nature of the earth and the heavens which of necessity changed man's concept of his own place in the universe. New knowledge led to new wealth, affecting the economic structure of the West. As bullion flowed in from the old and the new Indies and became available to the daring, the persevering, or the merely lucky, the new wealth became a new source of power. The discovery of heathen societies, unknown in Scripture, rekindled missionary fervor in the Catholic church, and at the same time a loosening of the social fabric resulted from migration and colonization brought on by the discoveries. These developments undermined the moral traditions and standards of communities which had long been relatively stable.

It was the task of the cosmographers to deal with the intellectual instability arising from the collision of new facts about the natural world with hypotheses built on old ones. Hitherto insoluble problems might well be reconsidered and give way to new approaches. Of the twenty-four kinds of ignorance defined in scholastic theology through ingenious distinctions, R. W. Southern has made a simplified generalization under two headings. One he calls "the ignorance of confined space"[1] which gave way before knowledge of the New World. This knowledge had to do battle with the second kind of ignorance, that of the "triumphant imagination." The cosmographers of Spain worked to

1. R. W. Southern, *Western Views of Islam in the Middle Ages* (Cambridge: Harvard University Press, 1962), pp. 14 ff.

1. A Cosmographer's World

confine imagination to the tenets of the Christian faith concerning the invisible, the immeasurable, the ultimate, and the forbidden as enshrined in the Old and New Testaments. They revived and applied the classical heritage of observation, geographical description, and scientific analysis to natural phenomena. Information from the New World resulted in a new curiosity and a new view of the Old World, and involved the indispensable link between the two worlds—the sea.

The science of navigation had to make great strides if its practitioners were to guarantee the safe passage of men and cargoes to the new lands. Officially, a teaching chair of cosmography attached to the Casa de Contratación existed from 1552 to the last nomination in 1707. But prior to the first date and subsequent to the last one, cosmographers were irregularly attached to the Casa and the Council of the Indies. Their duty was to aid in examining ship pilots and to make and check instruments and charts used for navigation. Both essential tasks were subject to demands for rapid expansion once the Indies grew to be Castile's busiest enterprise and Atlantic shipping (called *carrera de Indias*) became the lifeline of Hispano-American business. Corruption was ubiquitous, and because business brought profit, and illegitimate business brought faster and greater profit, there were attempts to get pilots' licenses without examination and to achieve accreditation of instruments by bribery, which promised profit to all concerned. Merchants wanted to move their goods, while some instrument makers wanted to keep their industry a monopoly.[2]

Scientific understanding increased with accumulated experience in navigation and discoveries of new lands. This in turn resulted in the parting of the ways between the theoreticians—astronomers and mathematicians—and the practical men, whose interests at times were diametrically opposed. There was seldom, however, a complete consistency in thought or action among either group, as was demonstrated by the changing of sides during the course of endless litigations. But the cosmographers found themselves a vital factor in the midst of the busiest port in Spain, Europe's most powerful realm, a setting which was the stage for brawls of harbor toughs as well as for intellectual combat among men who produced the most advanced theoretical speculation of the time. In their various ways, all of these people were trustees of the new knowledge, and they controlled the very door through which their society moved from the Old World to the New.

One of the cosmographers attached to the Casa de Contratación was Pedro de Medina, teacher, examiner of pilots, expert on charts and instruments, astronomer, chronicler, and moralist. He knew most of the approaches to knowledge that were open to men of his time. His long life provided an abundance of practical experience and exposed him to a variety of environments, from palace and castle to town house and quayside, which gave him the advantage of a variety of perspectives. Manuscripts from his hand are therefore worthy of consideration, not only in the spotlight of the history of science but in the diffuse light of Spain's imperial noon.

From a historical perspective, Pedro de Medina's most famous works are his textbooks on navigation, the *Arte de navegar* (1545) and the two entitled *Regimiento de navegación* (1552 and 1563), which carried his name throughout Europe and to America. In 1871 a copy of the Dutch translation of the *Arte* was found preserved in the ice on the track of William Barents's third voyage of 1596 from Spitsbergen to Novaya Zemlya.[3] The *Regimiento* passed the Strait of Magellan with Drake, and the tables of the *Arte* were consulted 144 years after publication by Captain Alonso de León

2. The instrument and chart market had become a monopoly of one family, the Gutiérrez, under the protection of the pilot major, Sebastian Cabot. This illegal monopoly was challenged in a law suit in 1543. The documents are in the Archivo General de Indias (henceforth, A.G.I.), Justicia Sección, legajo 1146, ramo 2. See also Ursula Lamb, "Science by Litigation: A Cosmographic Feud," *Terrae Incognitae* (Amsterdam) 1 (1969): 40–57.

3. David W. Waters, *The Art of Navigation in England in Elizabethan and Early Stuart Times*, p. 163, n. 2.

on an *entrada* in Texas, where he named a river after Medina, at Easter time in the year 1689.[4]

Commander D. W. Waters has noted both the *Arte* and the *Regimiento* in the possession of several distinguished foreigners, who presumably made use of them. Martin Frobisher had a copy of a *Regimiento* in 1576, when he was searching for the Northwest Passage. He had been taught by the famous Dr. Dee, who may have advised him of its value.[5] Drake had one of Medina's works aboard the Golden Hind in 1578, possibly seized when he captured the "pilot of China,"[6] Sanchez Colchero (*el moço*). The *Arte* is also supposed to have inspired a famous Italian didactic poem called *Nautica* by the mathematician Bernardino Baldi (1553–1617), and it was honored by other writers who copied parts of it.[7]

A summary account of the revisions, translations, and editions of the *Arte*[8] is another way to tell the story of its widespread and prolonged popularity. In 1552, seven years after the first edition, a condensed version of its most useful information was printed as *Regimiento de navegación*[9]—a brief and manageable textbook for pilots. This was followed by a revision, which was newly licensed and contained twenty new *avisos* in the second part, the *Regimiento* of 1563.[10] The current authority on the bibliography of the *Arte* is Vice Admiral Julio F. Guillén of the Museo Naval in Madrid, who in 1943 published a list of books on navigation written in Spain, including all known editions of Medina's works.[11] He has since corrected and updated the list, but has not yet republished it. There seems to be at least one ghost edition (1573), repeatedly recorded but never described. Guillén lists over twenty foreign editions, which gives some idea of the importance of the work: the last French edition appeared in 1633, eighty-eight years after the first printing of the book in Valladolid. The first French translation had been made by Nicolas Nicolai (eventually "primier" cosmographer of Henry II) and published in 1554. The same year saw the first of three Italian editions by "Fra Vicenzo Palatino da Corsula, bacilier" published in Venice.

The third and last Italian edition was issued in 1609. There were four Flemish editions, of which the first was by Michiel Coignet, who in 1581 published his own work on navigation, preceded by Medina's *Arte*. For English readers, John Frampton, one-time victim of the Seville Inquisition but nonetheless an admirer of the pilot major's office, brought out the first of two English editions in 1581. The second was in 1595.

Many contemporaries knew Pedro de Medina as the author of other works, among them the *Libro de grandezas y cosas memorables de España* (1548), a first "guide to imperial Spain" as it has been called, truly a sixteenth-century Baedecker.[12] In 1555 appeared his moral tract, the *Diálogo de la verdad*. This was widely read and saw thirteen Spanish editions in the sixteenth century. Dedicated to an official of the government of Peru, Pedro de la Gasca, first president of the Audiencia, it is also mentioned in a sixteenth-century book list in

4. Edwin P. Arneson, "The Early Art of Terrestrial Measurement and Its Practice in Texas," *Southwestern Historical Quarterly* 29 (1925): 84.

5. Waters, *Art of Navigation in England*, p. 144.

6. Ibid., p. 353.

7. A. González Palencia, ed., *Obras de Pedro de Medina*, p. xix.

8. *Arte de nauegar en que se contienen todas las Reglas, Declaraciones, Secretos, y Auisos, q̃ a la buena nauegaciõ son necessarios, y se deuē saber, hecha por el maestro Pedro de Medina* (Valladolid: Francisco Fernández de Cordova, 1545).

9. *Regimiento de Navegación: En que se cõtienen las reglas, declaraciones y auisos del libro del arte de nauegar* (Seville: Juan Canalla, 1552).

10. *Regimiẽto de nauegaciõ: Contiene las cosas que los pilotos hã de saber para bien nauegar; y los remedios y auisos que hã de tener para los peligros que nauegando les pueden suceder* (Seville: Simón Carpintero, 1563). Reprinted in facsimile with transcription, 2 vols. (Madrid: Instituto de España, 1964).

11. Guillén, *Europa aprendió a navegar en libros españoles*.

12. A. González Palencia, *La primera guía de la España imperial* (Madrid, 1940). An extension of this study constitutes the introduction to the same author's edition of Medina's works, in the series Clásicos Españoles (Madrid, 1944).

Mexico.[13] When these two works were reprinted in 1944, in an edition by Angel González Palencia of the Royal Academy of History in Madrid, Medina received modern recognition as an outstanding Renaissance author: his books were collected in volume one of a series called Clásicos Españoles. Finally, he wrote a chronicle of the house of Medina Sidonia for the patroness of his later years, Doña Leonor Manrique, countess of Niebla. This was the history of twelve successive generations of the heads of the Guzmán family, who were Señores of San Lúcar de Barrameda and of many other properties, mainly in Andalusia. Carefully assembled, it was made up like a garland embroidering the author's loyalty to the house. The *Crónica* was published in 1861 as part one of volume 39 in the *Colección de documentos inéditos para la historia de España*.

From this brief account of Pedro de Medina's published works, it is clear that an unpublished manuscript by him promises surprises in subject matter, forms of expression, and ideas, since he was a versatile and far-ranging author. The chance finding of a manuscript of the *Libro de cosmographía* in the collection of the Bodleian Library at Oxford in 1959 led to the eventual discovery that there are preserved four cosmographies by Medina: the *Libro de cosmographía* (1538), the *Coloquio* (1543), the *Suma* of Madrid (1550), and the *Suma* of Seville (1561). All but one of these have remained in manuscript, and the exception, the *Suma de cosmographía* of Seville, was not published until 1947. The four works were compiled during more than a quarter-century, a period in which the teaching of cosmography to mariners became an established academic discipline. Because the body of information collected in these manuscripts remains fairly constant, the text of one may serve as the basis for the consideration of all.

Each of Medina's dissertations stands at a certain point in the progression from ignorance to knowledge which makes up the history of nautical science. His contribution to knowledge must be assessed with

reference to two distinct aspects of the discipline, namely, science as a body of true statements about nature and science as a method of investigation and analysis.[14] The failure to make that distinction has distorted the record of the history of nautical science in the case of Medina; his lack of descriptions of new phenomena or hypotheses has overshadowed his contribution to teaching methods and concepts which created the possibilities of progress. It is particularly with respect to the application of astronomical observation in finding position at sea and of projecting a course for a ship that Medina designed a modern teaching program. In a text probably written by him in 1544, one reads that the art of navigation is: "to know how to conduct a ship from one place to another, and as there are no roads upon the seas they are taken from the sky. For this it is necessary to take the altitude of the sun, and likewise that [the altitude of] the [North] Pole must be known; [one must know as well] the compass, the calendar of the moon and tides, and other such things and their rules."[15] The job of the navigator on the high seas out of sight of land was to fix his position by stellar sights and to determine his course.

The earliest use of the stars in navigation was probably as a rough sort of compass, but finding position by the stars was quite another task. It presumed knowledge of the spherical shape of the earth and the lines of latitude (i.e., breadth upon the globe represented as degrees of "climates" since ancient times). Lines of longitude were imagined to form a grid with latitudes upon the surface of the globe, and a ship's position was theoretically to be fixed with respect to

13. Irving Leonard, *Books of the Brave*, p. 203 mentions six copies. The *Grandezas* also occurs in a library list of the house of Fugger; see Paul Lehmann, *Eine Geschichte der Fugger Bibliothek* (Tübingen, 1956), p. 243.

14. D. E. Gershenson and D. A. Greenberg, "How Old Is Science?" *Columbia University Forum* 7 (1964): 27.

15. *Coloquio sobre las dos graduaciones . . .* , in C. Fernández Duro, *Disquisiciones náuticas*, vol. 6, *Arca de Noé*, p. 513. Concerning this text, see Lamb, "Science by Litigation: A Cosmographic Feud," pp. 49-50.

these lines. The latitude of a position was determined by instrumental observation of the "altitude" or angular distance above or below the equator of the sun or of the polestar as it would be seen from the center of the earth. Ports, land-falls, and landmarks were entered by altitudes on the maritime charts. Longitudes giving the east and west distance from a prime meridian run from pole to pole north to south on the surface of the globe. One of the difficulties was that a prime meridian could not be fixed, and another was uncertainty concerning the distances between longitudes at various latitudes. Although Francisco Falero and Martín Cortés, both authors of texts on the sphere, describe longitude graphically as the rounded edges of a slice of an orange, the distances between such lines could neither be measured at sea nor projected on a chart for lack of mathematical tools, information about the globe, and instruments such as clocks. Chart-makers relied upon the older method of locating coastal points by the distance and direction between them in the style of the portolan charts, which had been developed in the Mediterranean.

Readers of English are fortunate in having available recent works written with clarity and elegance on the history of nautical science and astronomical sailing. Most recent is the work by David W. Waters, *The Art of Navigation in England in Elizabethan and Early Stuart Times*, which deals with the early development; chapter 2 gives a concise and well-illustrated presentation of the Iberian literature. Among the works of E. G. R. Taylor, *The Haven-Finding Art* and *The Geometrical Seaman* cover the theory and application of instruments used in astronomical navigation. A brief explanatory survey was written by Lawrence C. Wroth, called *The Way of a Ship*, and a comprehensive survey is by Salvador García Franco, entitled *Historia del arte y ciencia de navegar* (especially useful is volume 1). The recent work of other Spanish and Portuguese scholars and international contributions to the field are reviewed in an article by Beaujouan and Poulle, "Les origines de la navigation astronomique aux XIVe et XVe siècles." In addition to the text of papers given at the first Colloque International d'Histoire Maritime (in which that article appeared), the later volumes of the series report discussions by many authors currently writing on the subject. The only close scientific analysis ever made of Medina's work, though quite old, is still considered valid—the now very rare *Crítica* by Rafael Pardo de Figueroa, published in Cadiz in 1867. It deals specifically with the second *Regimiento de navegación* and includes "a glance at the *Arte de navegar* (1545) and the *Suma de cosmographía* (1561)."[16]

Modern appraisal of Medina's work has proceeded from the direction of either literary studies or history of science. These two approaches serve better to illustrate the state of their respective disciplines than they do to describe the author and his ambience. An attempt to reconstruct the world and the universe of Pedro de Medina over the span of his life has become a more attractive enterprise since the two unknown cosmographic manuscripts have turned up recently. In our time the coexistence of a multiplicity of worlds is taken for granted. Included are the scientific constructs not only of galactic and atomic dimensions, but also the worlds recovered by archaeologists from vanished times and by anthropologists from distant places. The navigator's universe is that forged by a group of people whose world required that they reach out beyond their shores into the unknown oceans and who had to choose their scientific, practical, and moral guidelines to sustain them on their venture.

16. Commander Waters is working on a study of the scientific contribution made by Medina in his nautical works which he bases upon the copy of the *Coloquio* of the Taylor Collection. This will soon be published.

In the year 1538, Pedro de Medina had gathered together the materials he wanted to submit to the Royal Council of the Indies in support of his manuscript of a *Libro de cosmographía*. The earliest contemporary mention of Medina's *Libro* appears in a royal order which was signed in Toledo on 20 December 1538. It reads in part:

> in as much as on your behalf, Pedro de Medina, cleric, citizen of Seville, I have been given a report that you make maritime charts, regiments, astrolabes, quadrants, compasses and cross-staffs, and all other instruments necessary for navigation to the Indies, moreover, dissatisfied [with the model on the market] you submitted a new regiment of the sun and the North Star and a *Libro de cosmographía*, which you have written, [along with an application for a license to make such charts and instruments for sale in Seville] . . . we hereby inform our officials in Seville that this license is granted. . . .[1]

The only reservation appended was the usual one that, prior to their sale, all instruments and charts were to be subject to the approval of the pilot major and cosmographers of the Casa de Contratación.

Soon after the date of the permit, Pedro de Medina, the newly licensed "cosmographer to his majesty" Charles I (of Spain, V of the Empire), settled down in Seville to begin an active career which was to last almost thirty years, until his death in 1567. By 1538, the year of the *Libro de cosmographía*, the Ptolemaic universe of concentrically moving spheres had won renewed authority through widespread publication of the *Almagest*.[2] This ancient construction of elements and stars was made to move within the empyrean of the Middle Ages, a Christian "envelope" of infinite divine power and will. Now, however, it had to enclose a new and altered map of land and oceans. One achievement of the "Age of Reconnaissance"[3] concerning the globe was the use of observed regularities in the stellar sky for fixing places upon the earth and for charting a course upon the ocean. The movements of sun, moon, and constellations, seen against the fixed stars, were revealed as reliable guides to men who ventured beyond sight of a coast. Over the centuries and throughout the world, a variety of practices had grown up among travelers and navigators making use of astronomical observations and natural aids, such as the property which a lodestone can give an iron needle so that it will point north. But these observations and aids had been developed in widely separate regions and only for special routes. Europeans had learned from the desert travelers and from seamen of the Indian Ocean, while they themselves had developed charts and pilot books, or rutters, for the Mediterranean and northern waters.

The imperial enterprises of Portugal and Spain which followed the discoveries and the rivalry which grew up between these powers called for a systematic collation of information on the position of the new lands, their outline and situation in the oceans, and the routes there and back. This above all else needed to be known before the grants to "discover"[4] could be implemented.

How to fix a place on the globe by methods which were theoretically valid and how to lay a course across the oceans to a distant landfall were two vital problems of the day. Concerning the first—a spot on the map—agreement was reached on the use of astronomical observations and mathematical tools. As for the second —how to get to a given place—every prince had to look to his own resources, new techniques, good ships, good men, and hardy spirits. This made for a saga of spectacular accomplishments, and such was the age of the

---

1. José Toribio Medina, *Biblioteca hispano-americana*, 1: 193-4.

2. For editions of Ptolemy, see Carlos Sanz, *Geografía de Ptolomeo* (Madrid, 1959).

3. The term is given as the title of a book on this era by J. H. Parry (Cleveland: World, 1963).

4. Samuel Eliot Morison, *Portuguese Voyages to America in the Fifteenth Century*. Morison's concise introduction includes a bibliographical discussion of the meanings of the word "discover," which might be equivalent to "explore" and, one may add, in the Spanish sense, under some conditions, to "claim."

2. The Author: El Maestro Pedro de Medina

great discoveries. From each of the voyages theoreticians gained new information, sailors new experience, businessmen increased incentive, and governors expanded ambition. One common cause among these frequently discordant groups was nowhere better reflected than in the pedagogic literature coming from the naval school of Seville, of which the *Libro de cosmographía* is a forerunner. But Medina's cosmographic works were not written as recollections in tranquillity, and the serenity of their prose raises them in one's estimation when they are seen as a counterweight to the noisy strife and disputes which involved the author, an embattled teacher of pilots in that port city of Seville.

The *Libro de cosmographía* of 1538, read by itself, stands in danger of appearing inferior to contemporary scientific literature because its scientific information was built upon reports rather than invention, and because its questions appear childish, its answers cumbersome. The work as a whole might seem unrepresentative of its author, or a minor effort, because it was never published. But the *Libro*, which started Pedro de Medina upon his public career as a cosmographer, was not the work of a beginner. It was written by a mature and experienced scholar who at the age of forty-five sought to embark on a new career and to obtain a license as a professional cosmographer. From the first day of his activity in Seville, Medina ranked among the senior staff of the scientific office of the pilot major, which controlled the pilots and masters and the scientific and nautical equipment employed in the *carrera de Indias*. His *Libro*, as well as his regiments submitted in 1538, were the works of a teacher, not of an apprentice. In the *Libro*, Medina presented a textbook on the nature of the universe as applicable to the art of navigation, which, though it appears weighted on the theoretical side, was primarily concerned with the use of scientific information in the practice of navigation.

But Medina's life had not been solely directed to the study of nautical science and the writing of the *Libro*, nor did he abandon his other interests in history, geography, or moral philosophy while he wrote that text. The *Libro* of 1538 is a work which sums up the entire past and foreshadows Medina's productive years as cosmographer, teacher, and author. It stands near the midpoint of his life, at the opening of his public career, as the earliest extant text from his pen. The *Libro* serves admirably as an introduction to Pedro de Medina's total accomplishment, to the circles in which he moved, to the Spain in which he lived, and to the view of the world and the universe which he commanded.

Medina spent his first forty-five years not only filling a great reservoir of knowledge and skill which was to serve him well in his later work, but also living an active life of service as companion, librarian, and preceptor in a great Spanish house. He was born in 1493, to parents who belonged to the household of the dukes of Medina Sidonia. Two towns, Medina and Seville, claim him as their son, but neither can be confirmed as his birthplace. The facts in favor of Medina are his name and the mention of that place in later accounts of his childhood. He tells how, as a boy, he witnessed the visit there of a Dr. Tello, who was arbitrating between Don Pedro Girón and his mother-in-law, Doña Leonor, wife of the duke of Medina Sidonia, in one of those frequent contests which occur in noble families. Medina's Sevillian origin was first suggested by Nicolás Antonio in his *Biblioteca hispana nova*, thus: "Petrus de Medina, domo hispalensis." But the royal cedula which characterizes him as *vecino de la cibdad de Sevilla* refers to his residence, not necessarily to his origin. Luís Toro Buiza, who searched parochial records at Seville and Medina, found none concerning Medina's birth.[5] Nothing further is known about his family, and Pedro himself first appears in written records as tutor to the eldest son in the house of Alfonso de Guzmán, sixth duke of Medina Sidonia. This heir, Juan Claros, born in 1520, was the first of five children,

5. Toro Buiza, "Notas biográficas de Pedro de Medina," in *Revista de estudios hispánicos*, 2:35.

three boys and two girls, and the only male to survive childhood.

The position of tutor in a noble house was one of high prestige in this period,[6] and a family such as the Guzmáns could be expected to demand for their son a tutor of superior education and dignified bearing. Such qualities were undoubtedly his when Medina took charge of the education of Juan Claros. He must by then have graduated from a university, and that of Seville is presumed the most likely by González Palencia. Medina must have taken holy orders by 1538, since the cedula mentioned above addresses him as *clerigo*. No further information concerning that phase of his life is known except that his patroness, Doña Ana de Aragón, was exceptionally pious and might well have exerted great influence over the tutor of her son.

Medina, in his own words, had dedicated himself to the study of nautical science at the age of twenty-five.[7] The University of Seville, however, is not known to have had a faculty of science during the years when he might have been expected to have attended its courses, although Seville had been included in the university reforms of the Catholic kings.[8] If, therefore, Medina did not learn mathematics and astronomy at the university, and since no record of formal enrollment elsewhere has turned up, it is likely that he was self-taught, having acquired his profound knowledge from the library of the dukes of Medina Sidonia, a distinguished establishment well known in intellectual circles of the day. In his last published work, Medina mentions having gone to sea just long enough to observe what he needed to know in order to understand the use of nautical instruments in practice. This voyage included a great storm, which he vividly recalled more than three decades afterward. As there is no later record of this event, it must also antedate the *Libro* of 1538. One may gather from his other works that Medina accompanied the duke, Don Alfonso, on several journeys through parts of Spain, spending fairly long periods at Valladolid, Burgos, the monastery of Guadalupe, Malaga, Cadiz, and at San Lúcar de Barrameda, which was the family seat and probably the main base for the schooling of young Juan Claros. Barrántes Maldonado describes with admiration the well-rounded classical education which Medina's student displayed, thus by implication reflecting upon the maestro's high standards as a tutor.[9] Medina was aware of the extraordinary intellectual activity in Spain during his time, and his descriptions of libraries and bookmarkets where he encountered them have been a source of information about the Spanish Renaissance to historians ever since.

There is no doubt that his mind quickened when presented with problems of navigation at sea. Matters concerning the coasts and seas, from fortifications of the coasts to fisheries and shipbuilding, are among the rare recorded memories of his early years. He describes fishing for tuna at the town of Conil in 1505 when he was twelve years old, as well as a pirate raid in the same town on 21 July 1515.[10] For anyone with an interest in the sea, San Lúcar was ideally situated, standing as it did at the very door of the trade routes to the empires in the East and the West. Medina tells in his *Grandezas* of the impressions made upon him and his contemporaries at the return of the *Victoria*, the only one of Magellan's ships to come back under the command of Juan Sebastián del Cano in 1522, and he reports the information which he got from one of the voyagers, Diego de Sotomayor.[11] No one on the Andalusian coast could be left untouched by the thrill of such news nor fail to speculate upon its meaning. Pedro de Medina met the demands for interpreting events of this type by seeking two sorts of authorities: he went to the great

6. Caro Lynn, *A College Professor of the Renaissance: Lucio Marineo Siculo among the Spanish Humanists*, p. 78.

7. A.G.I., Justicia, legajo 1146, ramo 2.

8. A. M. Villa, *Resumen histórico de la Universidad de Sevilla*, chap. 1.

9. Alonso Barrántes Maldonado, *Ilustraciones de la Casa de Niebla*, 2:469.

10. Medina, *Crónica de los duques de Medina Sidonia*, pp. 276, 284-5.

11. Medina, *Grandezas*, p. 67b.

library for the guidance which the past might furnish to explain the new image of the universe, finding in astronomical and mathematical works the theoretical tools he needed, and he collected the latest news from returning crews. There was so much for the maestro to explain to young Juan Claros that one may well speculate whether some questions in Medina's first book, the *Libro*, may not directly reflect the lessons at San Lúcar. Why is the sky blue? What makes a whirl-wind? Medina fashioned his answers from the best available authorities, tradition and the latest observations, and his access to and command of both were remarkable.

In 1538 Medina accompanied his employers and his pupil, Juan Claros, to Toledo, where the Emperor Charles V had convoked a meeting of the Cortes. This was a celebrated occasion at which the grandees of Spain displayed their wealth and power, and their splendor reflected upon the members of the noble households. It was in Toledo that the emperor formally endorsed the succession of Don Alfonso de Guzmán to the dukedom of Medina Sidonia, having revoked an earlier grant made to the duke's elder brother, Don Alonso Pérez de Guzmán, who had been "certified" as weak-minded and impotent. Don Alfonso had subsequently married Doña Ana, the divorced wife of this elder brother, and she was the mother of Medina's pupil. She was a granddaughter of Ferdinand the Catholic, who was also the emperor's grandfather. Doña Ana's ambitious mother was responsible for her marriages, first to the imbecile duke, then heir to the title, and later to Don Alfonso when it was known that he would succeed.[12] This history was forgotten in the splendor of the occasion at Toledo, where the ducal succession was formalized and Medina's young charge, Don Juan Claros, was created Count of Niebla in a colorful ceremony.

The family relationship of the dukes of Medina Sidonia with the imperial house was reinforced by the political sympathies which they shared. In the revolt of the Comuneros (May 1520 to April 1521), the duke had been instrumental in keeping Seville and other southern towns, Granada, Córdoba, Gibraltar, Jerez de la Frontera, and Ecija, within the imperial camp. In return for his services the duke had received help in his family fight with Don Pedro Girón over his legitimate succession to the towns of Niebla, San Lúcar, and Huelva, all of which were granted to the duke by the emperor. With control over the family lands had come affluence, due to the demand from the Indies fleets for all kinds of supplies. The Guzmán fortune rose because the inflation which lowered the income from fixed rents and ruined many a northern estate found compensation in Seville in commercial outlets for produce at high prices, and because the Guzmáns had opportunities to partake in trade through agents in the wholesale supply of the fleets. The family fortune was said to have risen from 55,000 to 130,000 ducats annually over the sixteenth century.[13]

Pedro de Medina was not only a chronicler of the grandiose style of the public occasions at Toledo, but also the beneficiary of the intimate relationship of his employer with the imperial circle. He tells of the many visits of the duke of Bavaria (son-in-law of the emperor's brother) to the Guzmáns and the hunts those noblemen enjoyed together. The emperor, plagued with occasional attacks of gout, spent some of his leisure time listening to the fascinating accounts of cosmographic lore, "astrology, the sphere, theory of the planets, and matters concerning charts and globes," as we are told by Alonso de Santa Cruz, his cosmographer major and *contino*.[14] Santa Cruz describes how he was on occasion called to divert the emperor in this fashion, whiling away the hours of pain during the Cortes at Valladolid in 1537. No such account has come

12. Medina, *Crónica*, pp. 544 ff.

13. Antonio Dominguez Ortiz, *Orto y ocaso de Sevilla* (Seville, 1946), p. 54, quoting from Marineo Siculo.

14. At this time *contino* appears to have meant that the man so entitled was a resident-pensioner of the crown. Several other cosmographers were also *continos*. The passage is from Alonso de Santa Cruz, *Crónica de los reyes católicos*, I: xcvi; see also p. xiii.

down to us from Pedro de Medina for the following year, 1538, at Toledo, but the dedication of his *Libro de cosmographía* to the emperor was surely more than a gesture. Medina may have had hopes that Charles would see his work in its velvet binding. It is not known whether the emperor in fact ever held it in his hands.

The *Libro de cosmographía* reflects its author's experience with this world and his ease within it. It indicates high standards of erudition on his part and implies his obligation to offer that knowledge to the reader in the simplest truthful statement. The *Libro* is a reflection of Medina's situation in the great library at San Lúcar, where the bookish scent of old parchment mingled with the sea breezes which carried his dreams to Seville. There he longed to play an active role among the cosmographers, and the *Libro* was to be the preface to that career.

Medina's chance arrived once he left Toledo in 1538 and his part in the education of Juan Claros was ended. With his license as cosmographer, he was able to take up the profession for which he had prepared. Although he quit the sheltered world of the household of the dukes of Medina Sidonia, he never broke his connection with the family, and his pupil's wife, Doña Leonor Manrique, Countess of Niebla, became the protectress of his later years. With his move to Seville in 1538, he immediately entered upon a public career as licensed cosmographer.

The Seville of Medina's time was one of the glories of the Spanish golden age, in every respect as rich and varied in its culture, population, products, and ideas as enthusiastic contemporaries and historians have pictured it. The seaport properly speaking was the entire estuary of the Guadalquivir, enormously varying in width at different stages of the year and tucked behind the bar at San Lúcar. This was a treacherous obstacle for a fleet of sail trying to negotiate it at one tide, a maneuver intended to discourage pirate attacks upon single craft. In addition, there were two rock-piles, formerly the foundations of an ancient bridge, one of which was submerged at high water. An old pilot book, or rutter, of the thirteenth century has this to say:

> From the said mouth [of the Guadalquivir] as far as the city of Seville is 60 miles by the river. . . . From the said mouth 5 miles to the south-west is a rock called Peccato which shows above the water. And if you wish to enter the river, beware of the bank called Zizar [La Riza] to the west. And likewise of another bank to the east called Cantara, which is near the cape called Sirocca. In entering the said river by ship you must first take soundings, and note the buoys, and when the water comes in and the tide rises go by the course marked by the buoys. . . .[15]

As a port the town on the banks of the Guadalquivir, fifty-four miles from the sea, was inconvenient for all ships over one hundred tons and prohibitive for the larger galleys of the Indies trade, two hundred tons and over. Although its situation up river should have made it safe, there was an old inland road, well traveled for centuries, which connected Seville with Algeciras. From the Strait of Gibraltar, it passed through a large and unassimilated Moorish district; and even though history does not record an invasion from Africa, this does not mean that the possibility of one was absent from the minds of men in the sixteenth century.[16]

From San Lúcar de Barrameda at the mouth of the river to the Gradas, which are the cathedral terrace steps and center of Seville, news of ships and voyages was the stuff of everyday life. The estuary of the river between these towns constituted one large channel

15. From *Lo compasso da navigare* (1250), quoted in E. G. R. Taylor, *The Haven-Finding Art*, pp. 105–6. For contemporary directions, quite compatible with the text of 1250 cited by Taylor, see Juan Escalante de Mendoza, *Itinerario*, MS, Museo Naval, Madrid.

16. Abu-Abd-Alla-Mohamed-al-Edrisi, *Description de l'Afrique et de l'Espagne par Edrisi*, texte arabe publié pour la première foi d'après les manuscrits de Paris et d'Oxford avec une traduction des notes et un glossaire par R. Dozy et M. de Goeje (Leyden, 1886), chap. 2; for the state of war in the Mediterranean at that time, see Carlo M. Cipolla, *Guns, Sails, and Empires*, pp. 15, 16.

alive with shipping day and night. All incoming transatlantic traffic was supposed to travel up to Seville. Although the larger galleys could not negotiate the long and difficult channel, goods and passengers coming and going had to clear through the Casa de Contratación at Seville. Up river came the golden fleet, via San Lúcar and Bonanza, coasting by the large salt deposits which were stored in huge piles on the banks, skirting the bending Marismas, and passing by wharfs and markets to the city.[17] Though no shipyards for large craft could operate on the river, there were two repair yards along the estuary, at La Horcada and Borrego, in addition to the wharfs at San Lúcar.[18] The villages along this busy waterway were an ideal nursery for seamen as well as adventurers because the fabulous riches of the new world could only be reached by sea.

The hazards of oceanic travel did not seem to differ substantially from those of other travels to a landsman of the time. Great ships came to grief right at the bar of San Lúcar. Medina himself was witness to a salvage operation at the site in 1542.[19] The perils of travel did not necessarily depend upon distance, whether a sixteenth-century Spaniard ventured on the roads or the sea. The western Mediterranean was infested with pirates who scourged the Spanish shores from their hideouts in Africa, to wit the above-mentioned pirate raid on Conil in 1515 which Medina describes, still a bit breathless, in his *Crónica*.[20]

The city of Seville began to grow spectacularly only after 1530, but between that date and 1594 her population grew from about 50,000 to 90,000 people.[21] Seville by then had developed into a manufacturing town of the first order; it was a central market for the produce of all the surrounding countryside, an important banking center, an entrepôt for imports and exports, a gathering-point for businessmen of many nations, a powerful episcopal see with its own Holy Office of the Inquisition. As a flourishing cultural exchange it attracted many writers, artists, and tourists, as well as the famous ne'er-do-wells of Spain's classic fiction who became the heroes of Cervantes and Lope de Vega.[22]

Among this throng of active people no group was more essential to the maintenance of the splendor of Seville than the men of the sea. The Triana quarter of the town, situated on the right bank of the river Guadalquivir, across from the walls of old Seville, was home to many of them.

Within the walls of Seville itself, on the left bank, were rooted the powers of the faith, the law, and the government: the great cathedral with its famous Moorish tower, the Giralda, the Lonja or merchants' exchange, the Casa de la Moneda or mint, the Audiencia or city hall, and the Alcazar, palace of the Moorish kings, which since the reconquest had been taken over by the royal family. All of these buildings were in time to become worthy symbols of their functions. When the Indies fleet arrived, it would drop anchor near the Torre de Oro across the Triana to discharge its treasure. Heavy traffic by horse, mule, carriage, wagon, pedestrian, or any combination of them, crossed the river over the bridge which floated on seventeen wooden pontoons, necessitated by the sandy banks of the river. This bridge was a famous landmark, though not the stablest one. Barrantes Maldonado describes the disastrous collapse of the bridge in 1540 during the crossing of the duchess of Medina Sidonia with her

17. C. H. Haring, *Trade and Navigation between Spain and the Indies in the Time of the Hapsburgs*, p. 9; Huguette and Pierre Chaunu, *Séville et l'Atlantique, 1504–1650*, vol. 2, *Le traffic*, p. 294; Henri LaPeyre, *Une famille de marchands, les Ruiz*, p. 197, map.

18. Haring, *Trade and Navigation*, p. 272, speaks of the technique of careening a ship without beaching it by shifting the cargo. See also A. R. Usher, "Spanish Ships and Shipping in the Sixteenth and Seventeenth Centuries," in *Facts and Factors of Economic History*, pp. 189–213.

19. A.G.I., Contratación, legajo 5103, letters by Medina to the Casa: 11 January, 2 February, and 20 May, 1542.

20. *Crónica*, pp. 276, 284–85.

21. Ramón Carande, *Carlos V y sus banqueros: La vida económica de España en una fase de su hegemonía, 1516–1556*, p. 37.

22. Ruth Pike, "Seville in the Sixteenth Century," *Hispanic American Historical Review*, 41 (1961): 1–31.

train of mounted ladies and retainers, most of whom were thrown into the river. Fourteen of her young ladies were drowned, but the duchess was rescued.[23] The collapse of the bridge may have been due to one of the frequent crashes into the pontoons by oversize timber barges, which were often haphazardly steered down the swift waters to the shipyards by a single man with a long pole. Or the river might have been in flood stage, which frequently caused heavy inundations. No wonder then that Sevillians preferred crossing by boat and that the river was alive with craft of all descriptions plying back and forth for hire between the banks of Seville and Triana, after dark their lamps dancing like fireflies.

No record exists of Pedro de Medina's having his own house in Seville. The palace of the dukes of Medina Sidonia, on the "Cal[le] de las Armas," located in the barrio of San Blas, was an impressive oasis of privacy—including a fragrant orchard—in the busiest part of town. Although it is likely, it is not known whether the maestro ever lived there. In 1556, because of the housing shortage in Seville, a real estate speculator, Martín López de Aguilar, laid out at his own cost an entire system of paved streets on the grounds of the palace and proceeded to develop the properties to his own profit and that of the ducal exchequer. Only one address for Pedro de Medina in Seville is preserved. In 1543 he appears to have stayed at an inn called El Mesón de la Rabera on the Callejuela de la Mar.[24] This was a street giving on the Arenal, or promenade along the river, with its traffic, fair grounds, flea market, and rubbish heap (malbaratillo). It may be that the maestro's inn was located in a cul-de-sac, or in a building somewhat set back from the bustling traffic, if we accept the dictionary definition of "rabera" as the "hind part of anything."

The enervating climate, the humidity, high temperatures, and windless days of Seville, its situation low in the river valley with hardly any ground more than thirty feet above sea level, could be only somewhat ameliorated by the fresh water brought into town by an aqueduct serving many fountains, which Medina himself described with delight.[25] The other source of pleasure was the sweet fragrance of the famous citrus trees and flowers wafting over the plazas and patios of the city (concealing the effect of its multiple sources of pollution), which were praised by all visitors to the Andalusian capital. Yet the maestro was a partisan of changing residence with the seasons, like the herds of sheep which followed the sun.[26] His own migrations after 1539 can only be traced by chance mention in documents.[27] During all these years he was affiliated with the pilot major's office, and those connected with that establishment were by and large a mobile lot.

Upon settling down in Seville in 1539, Medina had to find a means of making a living. His license entitled him to manufacture and sell instruments and charts with the pilot major's approval. He was therefore subject to the rulings of this personage both for a license and for the opportunity to sell his wares. In theory, only three appointments concerning the technical aspects of navigation were recognized within the Casa de Contratación when Medina arrived: the pilot major, the cosmographer in charge of charts, and the cosmographer in charge of instruments. This division of responsibilities between cosmographers dated from the tenure of Alonso de Chaves and Diego Ribero in Sebastian Cabot's absence (1525-33). But the growing demands upon their services and the death of Ribero

23. Barrántes Maldonado, *Ilustraciones de la Casa de Niebla*, 2:468.

24. A.G.I., Justicia, legajo 1146, ramo 2.

25. *Grandezas*, chap. 43, p. 75.

26. Ibid., chap. 49, "Provincia de Extremadura," p. 102.

27. Toro Buiza has confirmed that Medina made his headquarters in Seville with only occasional interruptions between 1538 and 1545. He was at the wedding in San Lúcar of Don Juan Claros in 1551 and at his funeral in 1556, and he probably chose as his base soon thereafter the household of the Guzmáns in San Lúcar. That he died in the Parish of San Pedro in Seville, as stated by Navarrete, cannot be confirmed by documents. Toro Buiza, "Notas biográficas de Pedro de Medina," p. 35; Martín Fernández de Navarrete, *Biblioteca marítima española*, 2:585.

in 1535 resulted in a number of separate licenses being granted to individual cosmographers, cartographers, and instrument makers who were allowed to work for the Casa, and Medina was one of these.

With his cedula of 1538, Medina joined a staff of experts as a senior member, qualified by his age and experience. His rights, which he shared with a number of others, were a typical combination of straight pay for "services" rendered, license to collect specific fees where stipulated, and permission to charge the going market price for his instruments and charts upon approval from the pilot major. The crown policy was to avoid wherever possible any salary with a regular obligation on the royal treasury.[28] Preference was given to a multitude of small grants or reimbursements for expenses, plus cost of trips elsewhere to give advice—such as, perhaps, the salvage operation at San Lúcar—or to consult with juntas of experts, of which some records exist. The most rewarding aspect of this method of payment was frequently reserved to the future biographer or historian in search of the whereabouts of his subject, when his name turns up unexpectedly in a miscellaneous list of payments.[29] The various licensed cosmographers active in Seville can be found as court pensioners—continos—among them Santa Cruz and Mexía; or they might act in other capacities in the Casa, as did the same Pedro Mexía when he substituted for one of the judges.[30] Santa Cruz lived in the Calle Sierpes during Medina's time, but he was in complete retirement from the everyday business of the Casa. Francisco Falero, brother of Ruy, Magellan's cartographer, was a much traveled expert whose services were sought after in and out of the Casa. Medina was not granted a regular salary at the start, and later attempts by Sancho Gutiérrez to put him on the payroll appear to have been unsuccessful.[31]

Medina's interest and training predisposed him to the development of astronomical sailing and the making of charts and regiments. As for instruments, in 1538 he had submitted astrolabes, quadrants, ballestillas or cross-staffs, and the oldest instrument used at sea, the mariner's compass, which was also the one which presented the most recent challenge in the phenomenon of "variation." This discovery was subject to doubt, to confirmation, to erroneous interpretation, and to ingenious theory. Medina's stand on this issue appears in a law suit which was brought against him in 1543. It seems, however, that he had no ambition to manufacture instruments for sale, but only to supervise their production and conduct their examination. He did have ambitions to make and sell charts: this was the cosmographers' main business and it was a lucrative enterprise. To construct charts for the pilots, Medina needed to have access to the royal pattern map, or *Padrón Real*, which was the sole admissible master chart from which all those for sale to pilots had to be copied.

Although Medina's standing as a scholar and his social and scientific connections were impressive, or possibly because they were so, he had to face the rivalry of men of lesser status who at the same time were more securely rooted in the pilot major's office. Medina's entry into the world of the cosmographers was in part assured by the *Libro de cosmographia*. But the cedula of 1538 hinted that his license had been granted him because of criticism sent to the Royal Council of the Indies of the regiments currently sold in Seville, that is, his very license implied censure of the pilot major's office. Medina was therefore handicapped from the start. Sebastian Cabot did not permit him to consult the *Padrón* and then refused to license Medina's chart, copied from one on the market, because it did not agree with the *Padrón*. The ensuing argument became so

28. J. H. Elliott, *Imperial Spain, 1469-1716*, pp. 171-72.

29. Such a list, dated 1525, is to be found in Pulido Rubio, *El piloto mayor*, rev. ed. (Seville, 1950), app. 5.

30. On Mexía as "teniente de contador," see A.G.I., Justicia, legajo 1142, ramo 2. Juan de Mata Carriazo, editor of Pedro Mexía's *Crónica del Emperador Carlos V*, says Mexía was not contador of the Casa de Contratación (prologue, 1:xxiii).

31. See the foreword by Admiral Rafael Estrada to Medina's *Suma* of Seville (1561), p. 24.

lively and was so important for Medina's whole work that he could not retreat from his initial criticism of the office, and a complaint took shape which ultimately came before the bench.[32]

If the maestro's reception by the pilot major left something to be desired, the cosmographer major, Alonso de Chaves, welcomed him as a like-minded soul. Immediately upon his arrival, Medina was given an opportunity to function as examiner of charts and instruments and of pilots and masters. By implication, he became a teacher of pilots as well. The crown saw to it that he was granted access to the *Padrón* by the pilot major, because without knowledge of it he could not function. This provision was made in a cedula of 24 February 1539.[33] To Medina the teaching was congenial, and his work as a teacher was most fruitful. As much as anyone, he was responsible for building an academic curriculum to back up the practical experience of apprentice pilots.

When the maestro settled in Seville, the only routine established had been that of the examination. Pilots and masters had to prove that they were familiar with the route they proposed to sail, and with harbors and coasts; also that they were able to set a course by a chart, to steer by the compass, and to determine the latitude of their position. Besides this, they were required to keep track of their course and distance by dead reckoning. This presumed that they knew the sky-clock and were familiar with astronomical tables and their use; also that they understood the construction, principle, and use of instruments, the quadrant, astrolabe, cross-staff, and compass. Much of this knowledge they might acquire at sea under a competent pilot, but the demand for new pilots outstripped the available number of closely supervised apprenticeships at sea. The original rough outline for the examinations dates from 1527 and was amended in 1534 to include, in addition to pilots, the masters of ships, who might have to take command in an emergency. Initially, no fee was connected with the examinations. Since the law did not make it possible for people to live by teaching, that activity took place largely unaffected by the legal code. Like Medina, some were free to offer their teaching for a fee as private tutors, even though they also functioned as public examiners. This practice resulted in jealousy and corruption. Medina's enemies accused him and another veteran, Diego Sanchez Colchero, a well-known senior pilot of explorations and a shipowner, of charging money for instruction and of guaranteeing success in the exams, which they allegedly arranged by selling questions and by passing unworthy students. Some bribery was undoubtedly going on. The complaint was brought by Alejo Alvarez of Puerto de Ayamonte, who said he had paid eighteen ducats for his license.[34] What he was paid for his testimony is not known. Alonso de Chaves, legally responsible for the examinations and teaching, refused to sign the denunciation, but also refused to pass for license the candidates named in the complaint. No fine was recorded against Colchero or Medina, who had made his reputation over the years by publishing far and wide all questions and answers which might be of any use to the candidates.

No concessions granting a salary for teaching or examining were made to Medina and his colleagues. His attendance at examinations came under "services rendered," for which he periodically sought recompense from the council. Sebastian Cabot as pilot major regularly charged two ducats per examination, an "illegal" payment which by the latter part of the sixteenth century was tacitly accepted as a customary offering along with "guantes y gallinas" (gloves and chickens), the graduation payment to examiners in use

32. Lamb: "Science by Litigation: A Cosmographic Feud," pp. 40-57. Licenses and excerpted texts are given in J. T. Medina, *Biblioteca*, vol. 1, by index; also in his *El veneciano Sebastián Caboto*, vol. 1, passim; and in Pulido Rubio, *El piloto mayor* (1923), pp. 82-84, or rev. ed. (1950), pp. 487-89.

33. J. T. Medina, *Biblioteca*, 1:194. Toledo, 24 February 1539.

34. Haring, *Trade and Navigation*, pp. 375-77. This was Diego Sanchez Colchero, *el viejo*, who eventually became the owner of the *Trinidad* and sailed frequently to Santo Domingo.

throughout Spain and her empire. Richard Hakluyt describes the operation of the Casa de Contratación's teaching establishment in his plea for a like foundation in England, and relates the visit to the Casa by "Master Steuen Burrowes," who observed the formality of the granting of licenses and was presented with a "payre of parfumed gloues" worth five or six ducats.[35]

Spanish instruction in navigation was impressive. The systematic presentation of the subjects of nautical science in Medina's *Arte de navegar* and *Regimiento*s and in the books of Martín Cortés and Jerónimo de Chaves put Spanish training far ahead of the schooling available in other countries. Richard Hakluyt perceptively explained this in the "Epistle dedicative" of his *Divers Voyages touching the discoueries of America and the Ilands adiacent* (London, 1582). In this epistle he pleaded for the establishment of a navigational lectureship in London on the Spanish model. He mentions in his description the teaching and examination systems of the two Chaves and Pedro de Medina, "which writte learnedly of the art of nauigation."[36]

The *Libro* had shown Medina to be an erudite humanist and a knowledgeable scholar of cosmography. By the time the *Coloquio* was signed in 1543, he was the Maestro Medina in Seville, a title which he may have preferred in order to differentiate himself from two well-known contemporary Pedro de Medinas in Seville, one a secretary in a city office, the other an Indies merchant. The *Arte de navegar* marks the high point of Medina's chosen career, as indicated by his calling himself the "Maestro of the Arte de Navegar" in the *Regimiento*s and other subsequent books. There is no doubt that Medina's work as a teacher was respected by his peers, throughout Europe and over a long span of time. Yet he did not stand in the front rank of creative scientists, partly because he was not a practicing navigator, and also because his disciplined imagination was made rigid by his classical training. His denial of the possibility of the declination of the compass, which had been observed on Atlantic voyages, was his most noted failure.

Medina's contribution as maestro therefore was not invention but explanation. His facts, "wrong" though some might prove to be, were never in the way of his message about what counts in science. He taught that one must observe truly, record faithfully, and interpret modestly. This was, in essence, a moral message. The dedication of the first translation of the *Arte* into French by Nicolas Nicolai, may be read in this sense:

Oh! Happy nation, you who deserve the praise of this world over whom nothing has prevailed, neither fear nor hunger and thirst, nor innumerable travail to keep you from circumnavigating the greater part of this world, over seas and oceans never suspected to exist, and through unknown lands of which no one had ever heard; *and all this only through faith and virtue....*[37]

A century later Father Dechales said a little less flamboyantly that the *Arte* contained "many good things which have by now become common knowledge, but which were much admired during his [Medina's] time."[38] Medina agreed with the prefatory remark of an earlier teacher, Abul Fedda, who said in his geography that "what cannot be totally known ought not to be totally neglected—for knowledge of a part is better than ignorance of the whole." The obstacles in the maestro's way to make "good things common knowledge" resulted in a turbulent career, full of litigation, official inquiries, and eventual legislation which reflected the importance of nautical science and of the cosmographers to Spain's imperial enterprise. The author of the *Libro de cosmographía* was a master among them.

35. Waters, *Art of Navigation in England*, app. 16, p. 542. See also *DNB*, 1st ser., s.v. "Steven Borough."

36. Waters, *Art of Navigation in England*, app. 16, p. 542.

37. Italics added. *L'art de naviguer* (Paris, 1554). Though virtue was recognized as an asset of science, it did not interfere with the behavior of the scientist Nicolai, who absconded with some English charts when he transferred his loyalties from the English court to France. Waters, *Art of Navigation in England*, p. 82, n. 1.

38. J. T. Medina, *Biblioteca*, 1:190–91.

Introduction

Pedro de Medina's four surviving cosmographies share a common body of knowledge, but they differ in origin, make up, and history.[1] The earliest in the date of writing was the latest to enter the roster of manuscript cosmographies known to collectors and scholars. The *Libro de cosmographía* is one of six Spanish texts in the Canonici Collection, now in the Division of Western Manuscripts of the Bodleian Library at Oxford. This collection was acquired by purchase in 1817 for the then large sum of £5,444, though Sir Edmund Craster, then Bodley's Librarian, writing in 1921, judged the collection "well bought."[2] It was acquired from a Venetian Jesuit, and its two thousand volumes are mainly of Italian provenance. The *Libro de cosmographía*, as a scientific work, is the exception among the six Spanish manuscripts, which otherwise fit the categories of classical and theological texts characteristic of the rest of the Canonici Collection.

How the manuscript got to Italy is unknown. Medina had contemporary admirers in Italy, but whether this manuscript played any role in spreading his fame must remain doubtful, in view of the popularity of his published work.[3] Even so, one expression of this admiration is worth quoting. It comes from the pen of a Venetian scholar, Daniele Barbaro (1513–71), who used the maestro's wind rose and quoted Medina as an authority: "come dice Pietro da Medina." He expressed his thanks to Medina and other authors in the following passage:

Io ho cercato imparare da ognuno ad ognuno che mi ha giovato resto debitore de infinite gratie e come dispensatore dei boni ricevuti da altri mi rendo . . . [I have sought to learn from everyone. To everyone who has helped me I remain a debtor of infinite thanks and a dispenser of the good rendered unto me by others . . .].[4]

The manuscript of the *Libro*, marked Canonici 243, is in quarto, written on sixty-two leaves of paper, bound in light blue velvet on wood boards and held together by metal clasps. It is written in a clear gothic hand, with the initial letter of each paragraph in red. The drawings which accompany the text are in four colors, and the margins are ruled in red. The work opens with a dedication and preface; the body of the text is presented in dialogue form. Eighty-two questions are asked by a university graduate (*licenciado*) and a pilot, and the answers are given by the cosmographer.

Pedro de Medina's second manuscript on cosmography, entitled *Coloquio de cosmographía*, was acquired by H. C. Taylor of New York to be part of his distinguished collection of nautical literature. It was placed on exhibit in the fall of 1963 at the Beinecke Rare Book Library of Yale University, where it will be on deposit. A quarto, it is written on eighty-two leaves of paper and bound in blind-stamped calf over paper boards, secured by ties. One page at the end with an accompanying drawing is written in a different hand. The text appears between red-ruled margins and there are side notes giving the subject matter of the paragraphs. The title page is followed by a table of contents listing ninety-three questions. The full title leading into the first paragraph of the text reads in translation as follows:

Colloquy between the magnificent Señor Comendador Pedro de Benavente and Pedro de Medina, master of the art of navigation, cosmographer of his majesty, in which are treated cosmography, the sphere, the altitude of the sun and of the North Star, hydrography which is navigation of the oceans as well as location of places on the earth and other useful things which it is good to know. . . .

1. The four cosmographies were first described briefly in U. Lamb, "The Cosmographies of Pedro de Medina," in *Homenaje a Rodriguez Moñino* (Madrid, 1966), pp. 297–303.

2. H. H. E. Craster, *The Western Manuscripts of the Bodleian Library*, p. 14.

3. That there were many editions of the *Arte* and the *Regimiento*s in the sixteenth century does not necessarily indicate a large number of books since editions were small; nevertheless, it does signify persistent interest.

4. Vassili Pavlovitch Zoubov, "Vitruve et ses commentateurs du XVIe siècle," in *La science au seizième siècle*, p. 85. The quotation is taken from a Russian edition of Daniele Barbaro's commentary on Vitruvius, on which Zoubov collaborated.

3. The Cosmographic Manuscripts by Pedro de Medina: A Commentary

The colophon gives the year 1543; tables of the altitude of the sun and the North Star, introduced by brief texts, follow. The *Coloquio* has no dedication or introduction; the text begins with the first question of the comendador which the maestro proceeds to answer.

In the year 1543 Pedro de Medina was already far advanced in a scientific dispute leading toward a formal law suit which was to split the pilots and cosmographers of Seville into two camps. The Comendador Pedro de Benavente, a knight of Santiago and *contino* of the crown,[5] was a figure of social prominence, interested in the fashionable new science which was necessary to the most profitable of modern enterprises, the Indies trade. Medina may have presented the *Coloquio* to the comendador at his request, or perhaps to influence a powerful supporter of his cause in the high councils at court.

The above-mentioned separate text and drawing on the last two pages of the *Coloquio*, executed by a different hand, are interesting for what they reveal of the situation in which the maestro found himself in 1543. They read in part:

It seems that the Maestro Pedro de Medina who wrote this book had intended to give only a brief description to the Señor Comendador Buenavente of the topics under consideration. From these he left out one of the most important and useful matters in speaking of the altitude of the North Star when he omitted to discuss and illustrate the number of leagues which each of the winds has along its rhumb, degree by degree; that is to say, each degree N or S has 17·5 leagues, and on any one wind it continues to have 17·5 leagues for the degree, as shown in the figure. Being well aware of this, a friend of Pedro de Medina [adds it] to the present work from which it was left out not for lack of knowledge [but it just so happened] that this book remained without this particular figure [*cuenta*] which is taken from a book of astrology and cosmography written in the Tuscan language by Conpani [*sic*] Patricio Florentino, master of mathematics. . . .[6]

It is interesting that the addition is made in a spirit of friendship and praise for the work; another point to note is that the passage identifies the comendador by a different version of his name—"Buenabente"—and that the drawing added comes from an Italian source. Perhaps Medina's anonymous friend was an Italian scholar. The omission of the table may not have been an oversight on Medina's part, but its addition indicates the interest taken by the comendador and his friends in providing precise and useful information for navigators.

From the *Libro* written in 1538 to the *Coloquio* of 1543, which adds the solar and lunar tables, a natural progression toward a growing professionalism is revealed in Medina's work. When he wrote the *Coloquio* he was an instructor in the nautical sciences at the Casa de Contratación. His famous *Arte de navegar*, which he was busy writing between 1540 and 1543 and of which parts were circulating among pilots and cosmographers at that time, is therefore contemporary with the *Coloquio*.[7]

The last two manuscript cosmographies of Medina, the *Suma de cosmographía* of Madrid (1550) and the *Suma* of Seville (1561), were written for a lay public and consequently did not stress the scientific apparatus required by navigators. The *Suma de cosmographía* of the Biblioteca Nacional at Madrid is of full folio size,

5. Benavente was admitted to the Order of Santiago in 1530. Archivo Histórico Nacional, Ordenes Militares, Expediente de Ingreso de la Orden de Santiago, no. 969. He is mentioned as recipient of money from the royal treasury as *contino de la casa real* in Archivo de Simancas, *Catálogo XVIII*, compiled by Concepción Alvarez Terán (Valladolid, 1949), on p. 15/16 in a list of the year 1525.

6. The figure added is similar to one included by Cortés in his *Arte de Navegar* of 1551 (Cortés's figure is described in Waters, *Art of Navigation in England*, p. 76); both show the distance in leagues (17½ leagues to the degree) and in degrees and minutes that had to be run on a wind to raise or lower latitude by 1°. The numbers appear substantially modified in the *Regimiento* of 1563. For such changes see Taylor, *The Haven-Finding Art*, p. 164.

7. A.G.I., Justicia, legajo 1146, ramo 2, Seville, 1544. "Pedro de Medina y otros cosmógrafos con Diego Gutiérrez, Piloto, sobre la orden que se ha de guardar en la navegación y ynstrumentos de ella." Fol. 100, Probança de Alonso de Chaves.

bound in parchment, written on vellum, and has only nineteen pages altogether. The first lines of each text and the drawings are elaborately gilded and painted in many colors. Each figure takes a full page and the brief accompanying text is limited to the page facing it. It is now agreed that the manuscript is "autographo," while the illustrations are certainly made by another hand, though quite obviously copied from Medina's material in other books. The drawings are much more skillfully done than those in the earlier cosmographies, though the numbers, which look less finished, may be the author's work. This manuscript has one distinguishing feature of particular importance: it contains material and figures which have no nautical significance whatever. Pasted upon the pages at the front and back are engravings, printed on paper and cut out, of five female figures, classically garbed sybils, standing in niches. Each of the three at the front has a title written above it: Time past; Time present; Time to come; and a line underneath. The two figures on the back page point out of their frames to a clock face drawn in ink and an accompanying verse.[8] On the third page of the manuscript is written "del Caño [canónigo] Mayans," also known as Gerónimo Ayanz, who was a member of Philip II's scientific academy and author of a scheme to solve the problem of longitude, which he submitted in 1610. On the verso of the same page appears the stamp of Pascual de Gayangos. It is not known how this work reached the Madrid Library, and the manuscript is not listed in the catalogue of the Gayangos Collection. It consists of ten figures with accompanying text and a schematic presentation of tides; its map of the Atlantic coasts, beautifully colored and detailed, but purely decorative, is based on the *Arte*. Although the work is not dated, the text suggests that it was written with the resources of a great library close at hand, probably at San Lúcar in the year 1550.[9]

At least one other manuscript on cosmography mentioned in the historical records has not been found. The *Libro de cuentas* in the domestic archives of the House of Medina Sidonia contains a receipt signed by Pedro de Medina on 30 June 1562 for books which he brought for the study of the duke and "for binding," and on 2 December 1563 Medina received twenty ducats for "six figures of outstanding things in this world."[10] The *Suma* of Madrid has ten figures, so this must be some other work. But it is quite likely that the Madrid *Suma* was written for a patron's library.

The *Suma* of Seville, dated 1561 from its colophon, is in the Columbina Library on deposit in the Cathedral Chapter of Seville. It is the only manuscript on cosmography by Medina ever published, albeit nearly four centuries after it was written. Upon the suggestion of Luís Toro Buiza (who had combed the archives for data on Medina which he published separately) and Ramón Carranza, the president of the Diputación Provincial of Seville, that organization published two hundred facsimile copies of the manuscript with a preface by Rafael Estrada, Admiral of the Spanish Navy, in 1947. The manuscript is written on a smooth, vellum-like paper, with red and gilded lettering at the paragraph openings. The work has a descriptive subtitle and prologue, followed by the twenty-eight chapters which comprise it,[11] while the *Suma* of Madrid has no subtitle, dedication, or preface. There is a charming sketch map of the Atlantic Ocean within a circle, representing the

8. The prints of the sybils have a mark, L. S., in one corner. They may have been made by a Flemish engraver, Lambert Sustris; see no. 1915 of François Bulliot, *Dictionaire des monogrammes* (Munich, 1833), 2:249. See also Ulrich Thieme and Felix Becker, *Allgemeines Lexikon der bildenden Künste von der Antike bis zur Gegenwart*, 37 vols. (Leipzig, 1907-50), 32:314-16, Sustris = Zutman? (born ca. 1515-20): Amsterdam, died 1568. I am indebted to Professor George Kubler for these references.

9. I am obliged for this date to M. Marcel Destombes who told me where to look in the drawing of the wheel of the moon. There the date 1550 clearly appears, though it does not show well on the microfilm. See Destombes, "Un astrolabe nautique de la Casa de Contratación (Seville, 1563)," *Revue d'Histoire des Sciences* 22 (1969): 58.

10. Luís Toro Buiza, "Notas biográficas de Pedro de Medina," pp. 31-35.

11. The *tabla de las cosas* has twenty-seven items, but the number 10 is used twice.

earth as the central one of the four elements. Two drawings of ships under full sail appear on page liii, facing a text about weather at sea. The manuscript has a solar declination table calculated for the latitude of central Spain and a perpetual lunar calendar. Its colors are vivid, and it is in an excellent state of preservation. This *Suma* of Seville has a foreword addressed to "the prudent reader" and so was obviously intended for publication. The table of contents lists the topics in the usual way, which in the text are described as "figura y declaración," so that all the drawings are made to take up a full page and the text is fitted in, together with chapter headings on separate pages, sometimes several to a single illustration.

The *Libro* and the *Coloquio* were written in dialogue form, because, as the author says in the *Libro*, "It seemed the most suitable style. . . ." The dialogue form had not lost its vogue from ancient times, and it was a favorite method for teaching purposes. The *Libro* employs two questioners who are answered by the maestro, one a pilot interested in technical matters affecting his art, and the other an educated layman, a *licenciado*. These categories neatly fitted the people who would see the work or be shown it for expert assessment in 1538 in the Council of the Indies. The *Coloquio* for the Comendador Benavente is less well organized and most of the additional material—it has ninety-three questions as compared to the eighty-two of the *Libro*—deals with technical nautical matters. Also, the *Coloquio* has attached to it two regiments; they apparently were submitted separately in 1538, together with the *Libro*, but not bound with it.[12] From the *Libro* to the *Coloquio* there is, therefore, a move toward the more comprehensive technical manual which Medina was developing until he presented his truly new concept for a textbook of navigation in the *Arte de navegar*.

When in later years the maestro returned to work on cosmography with his *Sumas*, he omitted strictly scientific navigational matters and produced a narrative text instead of a dialogue. But once again, his was a presentation with a difference. One might call the *Suma*s graphic representations, with even more data presented for visual perception than had been in earlier cosmographies. The *Suma* of Madrid (1550) goes furthest. It is a book of annotated pictures, not an illustrated text. Of greatest importance are the full page figures, and the texts are merely brief explanations. Here is a book to look at more than a text for reading.

To recapitulate, the development of cosmographic material in Medina's hands through his productive years can be seen from his prefatory matter and manner of presentation. The *Libro* was dedicated to the emperor and written for highly educated men. The *Coloquio* was written for a private person, the Comendador Benavente, in the dialogue form of informal instruction. The scientific works produced between the cosmographies were meant for the practicing pilots to whom the *Regimiento*s are addressed, "a los señores pilotos y maestros que usan el arte de la navegación de la mar." These works were dedicated to Philip II. The *Suma* of Seville was addressed to the "prudent reader" and the *Suma* of Madrid has no introductory matter at all. But in this latter work Medina finds room not only for "universal bodies" but also for diaphanous ones, namely angels, who had not appeared in any prior cosmographic work of his. Other clerical references make one suspect that his patron was either a churchman or a very pious soul. It can be seen therefore that the *Libro* contains the seeds of all the future works, directed as it was to the powerful and to the intellectual as well as to theorists and practitioners of navigation. In his *Libro*, Medina wanted to be useful to "those who wish to know something of this science [of cosmography] and especially to those who go to sea." By the time he wrote the *Suma* of Seville in 1561, he left out matters of purely technical interest to navigators and presented a conventional view of the universe. In the *Suma* of Madrid he had added to this picture "man, for whom this world is created and whose nature is

12. J. T. Medina, *Biblioteca*, 1:193: ". . . hicistes presentación de un nuevo regimiento de la altura del sol y del norte. . . ."

known to partake of the properties of all that is known, . . . reason like the light of the sun, ignorance like total darkness [*tinieblas*] and the humors of the four elements."

With some exceptions in the prologue of the *Suma* of Seville and text of the Madrid *Suma*, Medina's writing is characterized by sobriety and a total lack of fancy or extraneous allusions. Poetry to him was in what he observed, not in quotations from the poets. As a scientist, he wondered at the ordinary and had no need of the extraordinary. It was also characteristic of his style that he left his texts relatively bare of references and quotations. A man so learned did not need to prove his acquaintance with the scholar's library, and he did not choose to flatter his audience but to instruct them.

In all of the *Libro*, Medina cites directly only one authority, Ptolemy, who is mentioned as having stated that the earth, as the heaviest element, must be in the center of the universe. In the *Coloquio* four authorities are cited: Theodosius, in connection with the "nature of the sphere"; the "royal prophet" and Gregory for the location of hell in the middle of the earth; and lastly "Alfrango" (al-Farghani), the ninth-century author of a book on the elements of astronomy which gives estimates of the sizes of the heavenly bodies. There are some allusions to "the Greeks" without further identification. In the *Sumas* more authorities are listed, especially in the *Suma* of Madrid. In these works we are given a fair sample of the classical authors and of the Semitic and Catholic authorities usually met with in contemporary tracts on cosmography. But Medina puts his references where the information belongs, unlike many writers who simply appended long lists of authorities—the longer the better, and often irrelevant—to a preface or at the end of their books. The *Suma* of Seville gives eight references and the Madrid *Suma* even more, much like modern footnotes.

Classical sources, though cited for their traditional content, are not always the expected choice. We encounter Ptolemy's *Almagest* and *Physics*; Aristotle on climates and meteors; Hippocrates on the nature of elements; Pliny's *Natural History* for the origin of pearls; and Ovid, Lucan, and Virgil as authorities on the inhabitants of various climatic zones. The crystalline sphere is based on Genesis, Avicenna is given as an authority on the compass, al-Farghani for astronomy; Ambrose is quoted on the sun, Augustine on light, Basilius on water as rain in the air, Dionysius (of Alexandria) on fire, Isidore (of Seville) on the compass, Albertus Magnus on the sphere, the venerable Bede and Sacrobosco also on the sphere. The absence of Peuerbach as well as of Regiomontanus and other more modern authorities will be discussed later.

In general, Medina's picture of the universe is based upon the authorities whose work made up the common intellectual heritage of his time: Aristotelian physics, Ptolemaic astronomy, and Mosaic-Christian cosmogony. All these speak from the pages of his *Libro* as they do from the works of his contemporaries.

One may highlight the meaning of this frame of reference by formulating a contrast between the popular views of his time and ours. Most characteristic of Medina's epoch was the coexistence of certainty that there are ultimate answers and of uncertainty about current observations and trust in the senses and in instruments. Medina himself gives examples of the fallacy of vision.[13] In his time there still prevailed the popular assumption of the Middle Ages that history was truth or *gesta Dei*, because it was past, because it had really happened, because it was unique and distinguishable. Beyond this life, Genesis and the Apocalypse were equally true in explaining the origin and destiny of man. Medina's universe was limited and determined in time and content by the creation and the

13. Medina uses both the coin which changes size when seen in the water and the stick which appears broken. This irregularity proved the uncertainty of natural events which were mere accidents, "cosas acaecidas." It took until 1612 for the phenomenon of the broken stick to be used to prove regularities of the laws of refraction; Blaeu's *Light of Navigation* includes "a drawing of a tub of water with 'a staffe slope in it'" (Waters, *Art of Navigation in England*, p. 325).

promise of salvation. This can be put, in a deliberate exaggeration, in sharp contrast to the popular assumption of our day, that a statistic which can be repeatedly achieved is "truer" than any single event. Historical events are felt to be less true, that is, less scientific, just because they are unrepeatable and supposedly less knowable than current events.[14] Today we work primarily outward from the certainty of the present. In Medina's time, men thought inward from the certainty of origin and destiny contained in the created universe.

In view of this fact and of the deliberate scarcity of authorities in Medina's *Libro* and *Coloquio*, the use made of them takes on a special significance. It means that he did not regard the ancient sciences as true because they were historic, but because they represented observed data interpreted by mathematical method. He objected to the truth of absurd events represented as true because of their uniqueness, which supposedly guaranteed their direct link to creation. Medina thought personal observation of a phenomenon and a record obtained and well kept preferable to the description of a past event by a dead authority: when he used authority, it was to illustrate method, only rarely to illustrate fact. An exception, the origin of pearls, is of special interest because of the new fisheries.[15] Pearls were said to originate by the entry of drops of sweet water into the shells at high tide. Medina organized his chapters as demonstrations of events witnessed together with their explanation. This made him the friend of practical men who had enough training to follow his theoretical explanations.

The Maestro Medina was headed in the direction of scientific advance in his fidelity to observation and record of the new heavens and the new earth then being discovered. Compared in this sense to that of his contemporaries, Medina's work stands up well. The closest rivals to his *Libro* were its immediate predecessor, Francisco Falero's *Tratado del esphera y del arte del marear* . . . (1535), and Martín Cortés's *Breve compendio de la sphera y del arte de navegar* . . . (1551).[16]

The writers of these books were less discriminating in dealing with remote events than was Medina. From a nautical point of view, Cortés's sphere is quite cluttered. He says, "The sphere is all that is created by God in the universe which can be divided into three classes of beings: bodies or elements, spirits or angels, and partaking of the nature of both, men." Another division of bodies, according to Cortés, is into luminous ones, or stars; opaque ones, like the earth, or metals; and transparent ones, like water and air.[17] Falero said only that "the sphere is a composite of many parts below a surface." This does not help much, but it does no harm until he introduces the old "fifth essence."[18] His book was submitted for a printing license one year before Medina's *Libro*, and it was approved for publication on behalf of the crown by Juan de Salaya, *protomédico* and *catedrático* (holder of a university chair) of astrology at Salamanca. Salaya was a renowned mathematician concerned with "natural astrology" or

14. A. J. Ayer, *The Problem of Knowledge* (London, 1956), pp. 170 ff., especially p. 180.

15. Just at the time Medina wrote his *Libro* the pearl fisheries at Cubagua were legendary for the income they produced. An explanation of the origin of pearls was therefore very much in demand. Enrique Otte, in *Cedulario de la monarquía española relativo a la Isla de Cubagua, 1523–1550*, gives the documentation for the industry.

16. Falero, *Tratado del esphera y del arte de marear, con el regimiento de las alturas: con algunas reglas nuevamente escritas muy necesarias* (Seville, 1535); and Cortés, *Breve compendio de la sphera y del arte de navegar, con nuevos y instrumentos y reglas, exemplificado con muy subtiles demonstraciones* (Seville, 1551), MS. completed by 1545. José M. Millás Vallicrosa, in *Nuevos estudios sobre historia de la ciencia española* (Barcelona, 1960), p. 327, and also in chap. 20, "Náutica y cartografía en la España del siglo XVI," gives reference to a facsimile edition of Falero's *Tratado* (Zaragosa, 1945) with a foreword by J. Guillén.

17. Almost identical texts appear in the prologue of Medina's *Suma* of Seville but not in the text. Cortés dipped into fashionable humanist bibliography; he mentions Seneca and his tragedies, also Fulgentius, a discredited authority, and Boccaccio's *De la naturaleza y de los dioses* [*Genealogia deorum gentilium*; *De montibus, silvis, fontibus* . . .].

18. Medina in the *Suma*s equates the fifth essence with the empyrean, so called by "philosophers."

astronomy, as well as "judicial" astrology. This was the old tradition which tried to establish the connection between the courses of the stars on the one hand and the human condition and destiny on the other, a notion which still governed much scientific enterprise. Salaya was probably employed to cast horoscopes as well as to make medical prognoses by the stars.[19] That his judgment on tables constructed for the use of practicing pilots was sought by the crown shows the support of the scientific enterprise in Seville and the benefit of royal patronage to provide a link between experts.

Medina's abstinence from judicial astrology, an art with which he was surely conversant, in favor of strictly theological arguments wherever mathematics and observation did not serve, distinguish his cosmographic work from that of many of his contemporaries. His merit does not lie in a difference of belief, because there is no reason to assume that he held opinions contrary to the views of others on astrology, but in the setting out of his arguments. Because judicial astrology did not answer specific and predictable needs of navigators, it was omitted from a discourse meant for pilots. These are probably the greatest advances made by the maestro: the classification of information, the creation of a consistent terminology, and the development of a frame of reference for any given topic. One must extrapolate for this explanation, for which positive evidence appears in other texts, chiefly the legal suit in which Medina was engaged with his fellow cosmographers.[20]

Looking forward from Medina's position in history, one can see that though observation and record subjected to mathematical method advanced the development of nautical science, the keystone of Medina's system was still the world of Genesis. His argument against the moving earth, no matter against whom it may have been directed,[21] was still that the creation of man's universe was consistent with the account in Genesis of the six days' work (*Libro*, question 32). In addition to the limitation put upon scientific inquiry by the rigid framework of his universe, Medina allowed only one contradiction to cosmic law or observed fact—God's miraculous power. Just as in our day a human tragedy is frequently "explained" by reducing it to a statistic meaningless to those affected, so in Medina's time people resorted to explanation by miracle, which was safely anchored in common assumptions concerning the nature of God, his acts, and human destiny; the *maestro clérigo* lived in that ambience.[22] It was commonly assumed, for instance, that hell was both extremely hot and extremely cold because men stretched their imaginations to picture utmost suffering, which certainly included frost as well as fire. But it was known that these states could not coexist and that in the transition from one to the other there must be an area of comfort. This could not be allowed to exist in hell, so Medina, the measurer of all things, observer and faithful recorder of natural phenomena, interrupted his discourse on heat in the *Diálogo de la verdad*, and invoked God's miraculous power to suspend all known law in order to keep hell uncomfortable.[23]

Another limitation of Medina's thinking, which he shared with his contemporaries, was the result of his experience with human error. As an examiner and teacher of pilots, he was aware of much falsely recorded or badly observed data and of defective instruments

19. Salaya was a member of the College of San Bartolomé, renowned for its valuable collection of scientific books. He was famous for the first publication in Spanish of the perpetual almanac of Abraham Zacut. See Guy Beaujouan, "Science livresque et art nautique au XVe siècle," in *Les aspects internationaux de la découverte océanique*, p. 76.

20. A.G.I., Justicia, legajo 1146, ramo 2.

21. There were both ancient and modern arguments in favor of a moving earth. It is not unlikely that Medina had heard of the Copernican theory, but this may not have been foremost in his mind, because his rebuttal is along entirely traditional lines.

22. In this respect he shared a "mental set" of reconciling observation with expectation. On this see E. H. Gombrich, *Art and Illusion*, p. 60.

23. *Diálogo* [*Libro*] *de la verdad*, p. 418.

that constantly interfered with scientific explanation and hypotheses. As he put it in another text: "the fault is not in the instruments but lies in those who do not know how to make them or use them."[24] So Medina did not accept data proving the variation of the compass.[25] He dismissed the phenomenon because he was sure that the pilots simply did not make proper records, and who better than the teacher of pilots and masters was aware of their shortcomings! He lived to be confronted by incontrovertible proof of the variation, which he still refused to credit; as Edward Wright put it: "Peter of Medina laboureth greatly to proue that there is no variation of the compasse."[26] But in this error he had the company of experienced and famous contemporaries: Pedro Nuñez, the great Portuguese mathematician and earliest author of advanced nautical theories, and Pedro Sarmiento de Gamboa, cosmographer and explorer. One might call them the three doubting Pedros.[27]

The maestro of the *Libro* was a teacher of certainties. Medina did not choose to point to open questions of which he was well aware. There were so many of them! The *Libro* was meant to show what he knew and to prove that this knowledge would be useful and that it could be taught to others. With the *Libro* he obtained his license and soon joined the great teaching establishment, which has been called the first technical college, in Seville.

By the time he wrote the *Coloquio*, Medina had been working with the cosmographers for over three years as a participant in some of the work and as a witness to most of it. The informal presentation of cosmographical material in a dialogue between a teacher and a student reflects upon his experience as a teacher of pilots. The *Coloquio* is the true forerunner of the *Arte de navegar*, in which Medina completed the shaping of his material into a text for the navigator's use at sea. The *Coloquio* reflects, and the *Arte* is proof of, the enlightened atmosphere in the scientific office of the Casa de Contratación at Seville in the 1540s. A record of the *Arte*'s composition has come down to us which

is worth considering for the light it sheds on the *Coloquio*. In 1543 Medina had completed a compilation of all the material which he thought necessary to prepare a pilot for the theoretical work, or *arte*, of sailing a ship, which he distinguished from the practical work, or *oficio*. It is quite likely that this compilation was in fact the *Coloquio* or something very like it, because copies of his book were widely circulated during the same year. In his testimony Medina tells how, upon royal order to improve on this useful idea,[28] individual pages were distributed among the experts, the content matching their special competence. These men who "took them home" initialed their corrections and returned the material to Medina. The whole was then once more gone over by the author of the great compilation, the first *Arte de navegar*.

The only contribution specifically identified is that made by Alonso de Chaves, which appears in the testimony by his son Jerónimo, given before the judges of the Casa de Contratación in 1544.[29] He is reported to have suggested to Medina that all Latin words be translated into Castilian. This is interesting advice, because Chaves himself was the author of an early textbook for seamen, probably the earliest handbook for the practicing seaman and his instructor.[30] Ranging

24. *Coloquio sobre las dos graduaciones diferentes que las cartas de Indias tienen*, in C. Fernández Duro, *Disquisiciones náuticas*, vol. 6, *Arca de Noé*, p. 513.

25. A.G.I., Justicia, legajo 1146, ramo 2; J. T. Medina, *Sebastián Caboto*, vol. 1.

26. *Certaine Errors in Navigation, detected and corrected by Edward Wright* . . . (London, 1610), 13:131. In the *Regimiento* of 1563 Medina mentions that variation has been reported but remains unexplained, and by implication is not serviceable as a datum for calculation of longitude as suggested by others, or justification for manufacturing "corrected" compasses.

27. Julio Rey Pastor, *La ciencia y la técnica en el descubrimiento de América*, p. 85.

28. A.G.I., Justicia, legajo 1146, ramo 2.

29. Ibid.

30. C. Fernández Duro, *De algunas obras desconocidas de cosmografía y de navegación* (Madrid, 1895). These are MSS. in the collection of the Real Academia de la Historia in Madrid. The essay

over all matters connected with navigation, it was composed in four parts: *Quatri partitu*. . . . It included traditional rhymes chanted by mariners at their work or used as mnemonic devices for observing the signs of the weather in the sky or the presence of fish in the sea, as well as some astronomical lore, and other information on ships and naval warfare. Although Medina's text was meant for more sophisticated pilots, Alonso de Chaves advised him to dispense with classical tags and fancies. The *Coloquio* appears to have been written with this advice in mind.

Beyond this voluntary contribution by Alonso de Chaves, which is on record, one is led to speculate on the rest of the text. For instance, Medina knew Francisco Falero and regarded him as an expert witness in his behalf after Falero had testified on scientific matters in a legal action of 1543. Medina had copied entire paragraphs from Falero's *Tratado* of 1535 without specific acknowledgment. It is quite probable that, as one of the Casa's cosmographers, Falero suggested he do so, rather than put himself to the bother of amending Medina's "compilation." This was not uncommon at the time: there are identical passages in the *Arte* (1545) of Medina and the *Sphera* of Martín Cortés (1551), while the *Regimiento* (1563) of Medina recapitulates passages from Cortés. Falero himself has copied directly from Pedro Nuñez, and Sacrobosco's presentation of the *Almagest* was back of all of these works.[31] They all have the same images: spheres like onion skins, or like the white around a yolk of an egg, and so on. The similarity of these passages appears to have concerned very few authors. Diego Gutiérrez, not a writer himself, begrudged the information he had given (without extra pay), as did Sebastian Cabot, but they could not complain about stolen texts since they were not authors of books. Occasional identical passages seem in fact to have bothered people less than the priority of a type of work. In earlier times, when all writing was in Latin and the number of literate people was limited, oral tradition favored common treasure and no one cared who was the first saint eaten by lions.

Now everyone wanted to claim that he had written a "first" book, such as for instance one on the art of navigation. The medieval fear of novelty is expressed in the Spanish word *novedad*, which is always suspected of meaning bad news, but science led to good news and therefore these writers raced to be first in presenting anything new.[32] Falero claimed to have written the first book of instruction on navigation although Nuñez anticipated his theoretical discourse and Alonso de Chaves his technical chapters. Medina called his *Arte de navegar* the first—which it was by title if not by date (1545)—and Martín Cortés writing later than he (1551) presented his book as the first practical text of everything a navigator should know.

There are two aspects of this matter which are of particular interest. One is the degree of personal acrimony, jealousy, and libel which seemed the order of the day. Accusations were freely made and Medina was accused of some wrong in connection with most of his works: For instance, Cortés envied the priority of his *Arte*, and he was supposed to have copied his *Grandeza* from Ocampo.[33] But Medina had defenders as well as detractors. Alonso de Chaves, cosmographer major, spoke up for the *Arte*, saying it was the maestro who had put it all together,[34] and in a letter appended to the first edition of the *Regimiento* of 1552, in answer

gives excerpts from Chaves's textbook, *Quatri partitu en cosmographía pratica y por otro nombre llamado espeio de navegantes*. For a description, see Ursula Lamb, "The *Quatri Partitu* of Alonso de Chaves, an Interpretation," *Revista da Universidade de Coimbra* 24 (1969): 3-9.

31. José Gavira Martín, "La ciencia geográfica española siglo XVI, Martín Cortés, Martín Fernández de Enciso, Jerónimo de Chaves, Francisco Falero." *Publicaciones de la Sociedad Geográfica Nacional*, no. 7 (Madrid, 1931), pp. 7-15.

32. For the cult of novelty, see George Sarton, "The Quest for Truth: Scientific Progress During the Renaissance," in W. K. Ferguson et al., *The Renaissance, Six Essays* (New York: Harper & Row, Torchbooks, 1962), pp. 55-76.

33. Francisco Vindel, *Pedro de Medina y su Libro de grandezas*. Vindel says that Medina was the first to describe the city of Madrid (p. 5). See also J. T. Medina, *Biblioteca*, 1:232-36.

34. A.G.I., Justicia, legajo 1146, ramo 2.

to a dedication to him by Medina, Chaves wrote that "from such a tree only excellent fruit could come."[35]

Medina's last cosmographies like his first were written with the intention of telling all he knew about the universe. But they reflect the separation of general knowledge from the highly specialized subject of navigation by astronomy. Although the *Sumas*, written to be read rather than to be used, were embellished both in style and appearance with greater care than the *Libro* and *Coloquio*, they are still sober texts, meant not to amuse but to explain. This old-fashioned austerity stands in contrast to the cosmographic output of contemporaries of Medina's later years, who catered to a sensation-hungry public, or to mere curiosity about remote regions. These authors were untouched by the plea for disciplined thinking and critical observation which speaks from all Medina's work. The introduction to an English book, William Cuningham's *Cosmographical Glasse* (1559), may serve to point out what characterizes so many works in all the European languages at the time. In his preface Cuningham describes the benefits which we receive from cosmography to be:

> that she delivereth us from great and continuall travailes. For in a pleasante house, or warme study, she sheweth us the whole face of all the earthe, withall the corners of the same. And from this peregrination, thy wife with shedding salt tears, thy children with lamentations, nor thy frends with wordes shall dehort or persuade the. In travailing, thou shalt not be molested by the inclemency of the air, boysterous winds, stormy shores, haile, ice and snow coming to thy lodging. . . . In sailing thou shalt not dread pirates fear perils and great winds or have sick stomach through unwholesome smells. . . .

In this book, not only are the *arte* and *oficio* of the sailor merged with the craft of the surveyor and the job of a reporter, but the reader is treated as a passive witness to a great variety of matters, not as a participant rationally comprehending specific problems related to astronomical navigation. Medina's *Sumas* were also addressed primarily to readers and not to pilots, but they required study and consistent reference to a body of fact collected to explain the structure of the heavens and the earth of the new discoveries.

Medina's use of somewhat outmoded information in his *Sumas* shows these to be the works of an old man, no longer in control of the newest data. It is worth remembering, however, that Medina did not grow old alone: Spain's colonial enterprise had aged with him. Over the years of the mid-sixteenth century, it had matured to a point where purpose and method of dealing with the New World were as important as reconnaissance and communication. In fact, some of the earlier cartographic knowledge was disappearing and Medina's map in the *Suma* of Madrid shows no advance over that of the *Arte*. This development was not entirely due to censorship, but to other factors as well, among them the vast expansion of maritime traffic and the development of the convoy system, which needed few licensed pilots and brought little new data; the end of the great reconnoitering voyages in the Atlantic; and the interruptions of the teaching program in Seville.[36] Routine had overtaken exploration and broad-based curiosity was officially discouraged for reasons of security on the seas and with intent to limit foreign ambitions in the new possessions.[37]

The sum of Pedro de Medina's work on cosmography, in both printed form and manuscript, constitutes a major opus on nautical science. This production might well have filled a whole life, but after the *Arte de navegar* (1545), the maestro turned to other work: the *Libro de grandezas y cosas memorables de España* and the *Diálogo de la verdad*. It is worth noting that in these works Medina did not abandon his scientific orientation. Although cosmography was neither the source nor the essence of his concern in

35. *Regimiento de navegación*, copy in the Taylor Collection.

36. Pulido Rubio, *El piloto mayor*, rev. ed. (Seville, 1950), pp. 175 ff.

37. Alexander von Humboldt, *Kosmos*, 2:214.

either text, it is constantly reflected in both. Where in the *Libro de cosmographía* nonscientific analogies are made to serve cosmographic ends, the *Diálogo de la verdad* presumes acceptance of scientific navigation as a model for charting the moral void:

> And after having written the book on the art of navigation by which sailors govern themselves and their navigation, avoiding the perils of ignorance, it seemed to me that I should write another book for all of us who voyage over the tempestuous seas of this world, becalmed or through tempests, to arrive safely at our port of salvation.[38]

This reassuring picture of navigators sailing the unknown seas, calm in their knowledge rather than terrified by superstition and ignorance, was a new image to which the maestro had substantially contributed.

As Pedro de Medina advanced in age and gained in renown, he joined the expert councils in cosmographic debates as an authority rather than as an investigator. His contribution to science was, however, still substantial—not so much in scientific practice as in making clear what science could and should do. His testimony was sought on the *Padrón* of 1553/54, and he was called to court to give advice in the matter of the Philippines. This question was debated in two stages, in October 1566 and in July 1567. A junta of scientists was to decide whether the Philippines were on the Spanish or Portuguese side of the line of demarcation. Medina's last declaration as cosmographer of the Casa de Contratación in this debate is worth excerpting:

> I, Pedro de Medina, cosmographer, citizen of Seville, made a declaration in this city of Madrid on September 8 of the past year [1566] in which I said that the Philippines fall within the pledge which the emperor our lord who is in glory [Charles V] made to the king of Portugal. . . . But it was not my intention to declare, that because of this the Philippines were pledged to the King of Portugal, excepting that this be explicitly so stated in the text of the agreement. *Testifying to this has nothing to do with my profession,* nor have I seen the complete text of the agreement

which alone would enable me to deduce the intentions of the contracting parties. . . . And because I understand this to be the facts in the case I sign my name. Madrid, July 17, 1567. El Maestro Medina.[39]

His last service to science was to say: "testifying to this has nothing to do with my profession." He drew a clear distinction between a scientific hypothesis concerning the location of the Philippines and the validity of a diplomatic instrument, the pledge of one ruler to another concerning the claims of sovereignty over the islands. This famous and complicated controversy is of great interest for the state of knowledge of the Far East which it reveals. Medina's contribution, in view of his lack of specific new information, was to state that it was wrong to allow diplomacy to influence scientific hypotheses.

In addition to discussing the nature and the role of science, Medina promoted the idea of the need for its continued development. This he did, probably at some cost to himself, in the matter of the compass variation. Again his advice was weighted on the moral side of the question. In his *Regimiento* of 1563, he quotes Seneca as having said that "no time is too late to learn," and that "with one foot in the grave one would still wish to know." His last advice concerning the compass was a slight modification of his denial of variation. He was still willing to learn and advised that the pilots of the *carrera de Indias* be asked to carry two compasses, one corrected for variation and the other not, and that the pilots be required to keep records of their reading, "so that many useful things could be found out." Moreover Medina had long recognized the limitations put upon the knowledge and the work of any single man. In the preface to his *Grandezas* he wrote that he was

38. Medina, *Diálogo* [*Libro*] *de la verdad*, p. 262.

39. Italics added. For the complete text of Medina's statement, see Estrada's foreword to the *Suma* of Seville, facs. ed. (Seville, 1947), pp. 25–26. For a discussion of the junta of experts on the Philippines, which in addition to Medina consisted of Santa Cruz and Jerónimo de Chaves, see Santa Cruz, *Crónica de los reyes católicos*, 1:xlix.

certain that there were many errors because the subject matter was difficult and contained many details, but that except for the Bible—*las divinas letras*—there is nothing so well written that it does not need to be amended, corrected, and checked.

He put his last work, the second *Regimiento*, to press in "the house of Simon Carpintero, next to the church of San Pedro in the month of February 1563, the author being seventy years old [*y la edad del autor de setenta años*]." Perhaps his birthday was in February, and he celebrated it with that famous publication. The last documents concerning him are from the archives of the House of Medina Sidonia, and no further clue to the time or cause of his death, sometime about 1567, has been found.

Medina's *Libro de cosmographía* and later cosmographic work show the stimulation which scientific thought derived from the geographical discoveries. But the gap between theory and practice in nautical science is equally clear, due not so much to a difference in the temperaments of the practitioners as to the fact that application of theories was hampered by the lack of what now would be called development—that is, the modification of instruments for different uses or for use under different conditions than the customary ones. Also, gaps in theory had to be filled before nautical science could be more to professional pilots than intuition modified by experience.[40]

One point worth stressing and too frequently taken for granted is the fact that political lines of demarcation were drawn across the oceans by scientists. Papal decrees were validated by an astronomically determined line which separated the claims of Spain from those of Portugal in a part of the world still "to be discovered." Religious authority and legal experts, political advisers, and military technologists were joined by new arbiters in international disputes: the cosmographers. By their scientific consensus these astronomers, mathematicians, and geographers fixed a line on the globe, and held the spotlight as new umpires calling the score in the game of kings.

The *Libro de cosmographía* which Pedro de Medina wrote to introduce his contemporaries to their universe may serve as an introduction to the world of the cosmographers of the sixteenth century. This is a world faintly outlined and presenting many blank spaces, particularly in social history. Who were these international figures in an age of nationalism, these prestigious professionals, who, reduced to the democracy of common suffering aboard their ships, would walk ashore to deliberate with kings? These stargazers who would catch the movement in the heavens on a small scale or bronze ring and reduce the New World to lines on a chart? Anticipating Francis Bacon, who urged that nature be studied as the "Booke of God's Works,"[41] Pedro de Medina put into his book the world of the discoveries. Although we no longer see our universe in his *Libro de cosmographía*, we can find in it the world in which its author lived.

40. See Avelino Teixeiro da Mota, "L'art de naviguer en Méditerranée du XIIIe au XVIIe siècle et la création de la navigation astronomique dans les océans," in *Le navire et l'économie maritime*, ed. M. Mollat, Colloque international d'histoire maritime, 2d, 1957, especially the discussion by Chaunu, Godinho, and Poulle in the appendix, pp. 141 ff.

41. Bacon, *The Two Books of the . . . Aduancement of Learning* (London, 1605), fol. 6ᵛ.

LIBRO de COSMOGRAPHÍA
Facsimile

LIBRO DE COSMOGRAPHIA

En que se declara vna discripacion del
mundo, dirigido a la S. M. del enpe
rador don Carlos, nnestro señor.

fecho por pedro de medina cosmogra
es pho.

La experiencia Como madre delas co
ssas nos enseña quantos prouechos
e vtilidad entrelas sciencias q̃ escrip
tas son: la singular sciencia de cosmographia
nos trae e causa. Delos quales tres hallo muy
principales. El primero que por esta venimos
en mayor conoscimiento del muy alto poder e
sabiduria de dios porq̃ como trate delos cielos
yotras grandezas que enel mundo son. los cielos di
ze el Real propheta enarran la gloria de dios. E sant
gregorio las marauillosas obras que vemos cri
adas vestigio e señal son de nro criador. El segu
do que nos haze mas abiles para entender las
diuinas escripturas porq̃ enellas se contiene y
trata muchas vezes del vniuerso e partes dela
tierra abitada. El tercero que nos yntroduze e de
clara los libros dela natural philosophia asi co
mo de generacion e corrupcion. Decelo e mundo

e delos methauros y avn tanbien entendimiento
delos libros delos poetas Pues yo deseando al
go aprouechar alos que desta sciencia saber desean.
y especialmente dar avisos alos q la mar nauega
quise tomar carga allende demis fuerças. en ha
zer vna breue discripcion del mundo deq e es
cripto e copilado este libro Elo ordenado por
preguntas que hazen vn licenciado y vn piloto
a vn cosmographo porque me parecio por este
estilo sera mejor entendido lo que del se dira. E
por q libro que trata del mundo es justo que
a vra magestad. Como monarcha del mundo se
ofrezca. e yo asi lo presento e muy vmilmente su
plico tenga por bien mandan lo mirar con los o
jos que el señor miro la pequena offrenda de
sophia su sierva bien asi no acatando la po que
dad demi obra mas la grandeza dela voluntad cõ
que servir deseo :—

36

¶ Tabla delas preguntas q̃ s̃o ē este libro:

¶Quantas sonbras haze el sol alosq̃ abitan
enel mundo / p̃ᵃ.   25

¶En q̃ lugar se vee · el sol · y no haze sonbra al le
vante · ni al poniente · ni a niğuna delas otras par
tes / p̃ᵃ.   26

¶En q̃ lugar estan ygual mente apartados el
sol · y el norte / p̃ᵃ.   —   27

¶Porq̃ se cavsa el edipsi del sol mas en vn tie̊
po que en otro / p̃ᵃ   28

¶Que cosa es tienpo / p̃ᵃ   29

¶Que cosa es año / p̃ᵃ   30

¶Que cosa es mes / p̃ᵃ.   31

¶Que es semana / p̃ᵃ   32

¶Que es dia · / p̃ᵃ.   33

¶Que es ora / p̃ᵃ.   34

¶Porque vnos dias son grandes y otros pe
queños / p̃ᵃ.   35

¶Si por llegar senos el sol tenemos mayor dia porq̃
los q̃ está mas apartados del sol tiene̊ mayor dia / p̃ᵃ. 36

41

44

LIBRO DE COSMOGRAPHIA

En que se declara vna muy prouechosa dis
cripcion del mundo. es asaber delos cielos y
estrellas sol. e luna. yelementos. va ordenado
por preguntas. que hazen vn licenciado. y vn
pilo a vn cosmographo. Comiença del modo siguiente.

¶ LICENCIADO.

1 Pues avemos de tratar de Cosmographia pre
gunto q̃ es cosmographia. E dado se dize assi.

¶ COSMOGRAPHO.

Cosmographia es discripcion del mundo
dizese de cosmos. nonbre griego que
quiere dezir Mundo. e grapho. discrip
ció. Asi q̃ cosmographia es vna discripció del mu
do. E en esta discripcion ay. geographia e hidro
graphia. La geographia es discripcion dela tierra
ca si se dize de geos que es tierra. La hidrographia
es discripció dela mar. e dizese de hidros. que es agua
pues desta discripció de cielos y elementos de que

el mundo es conpuesto alpresente tratemos

LICENC̃ Que es mundo E porq̃ se dize
assi y que partes tiene. COSM-

MVndo es la vniuersidad delos hōbres
consta de cielos y tierra mar codelos o
ttos elementos · Es llamado mundo porq̃
sienpre esta en mouimiento que ninguna holgança
le es concedida Este mundo se diuide endos partes
orregiones es asaber Region celestial y Regiō elemē
tal Enestas dos Regiones ay Catorze cuerpos sin
ples los quales son El primero la tierra que es cen
tro delmundo Seca negra pesada egruesa El 2º
es el agua fria pessada diaphana El 3 es clare
vmedo subtil cligero· El 4 es el fuego caliente li
gero Subtil erresplandeciente Sobre estos quatro
elementos q̃ son principios de todas las cosas cōpues
tas de naturaleza/estan otros diez cuerpos q̃ se dizē orbes
celestes cada vno delos q̃les circūficie̅e alotro hasta pa
rar enel decimo q̃ a todos cerca y no es de otro cuerpo algū
o cercado El prīm̃ destos cielos es el orbe do esta laluna y
el q̃ esta sobreeste es mercurio El 3 venus El 4 el sol El

qnto mars el 6 jupiter el 7 saturno el 8 ٭ destā todas
las estrellas q̃ se dize firmamento y antes deste todos sō tras
parentes porq̃ vemos las estrellas las q̃ les no veriamos
si ellos nolo fuesen ٭ sobre este octauo esta el noueno cie
lo cristalino o tan luziente amanera de cristal ē el q̃l no
ay algūa estrella y este se llama primū mobile el deci
mo es el ynpireo q̃ atodos circū fiere y es yn mouible
e puta luz :||

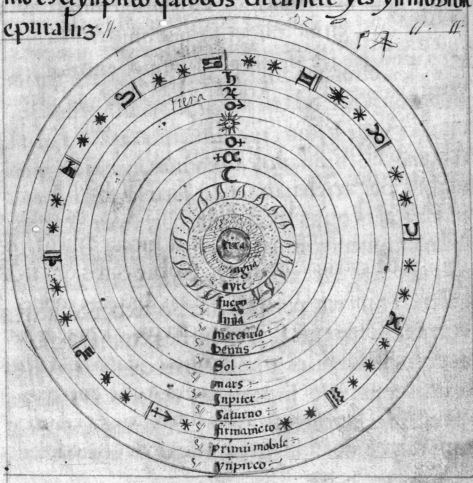

¶ LICENC̃ Las nueve speras si se mueven o no E si hazen todas vn mouimiento o difere̅tes ?

E.Stas nueue speras se mueven e̅dos maneras · vna es del primũ mobile de oriente en ocidente boluiendo otra ves ·a oriente · y las otras ocho ynferiores al co̅trario desta de ocidente en orie̅te Mas este moui̅ miento primero consu ynpetu eligereza arrebata todos los otros mouimientos y hazeles dar vnabu elta enderredor dela tierra envndia evna noche vna vez yeste sellama mouimiento rapto · o forcoso Algunos dizen quenosea de ymaginar q̃el primũ mobile por arrebatamiento forcoso oviolento · mueba alas otras speras ynferiores por Razo̅ q̃el mouimiento Rapto esfuerca oviolencia ·y esta nopuede aver enel cielo porque alli noay Re sistencia · tanbien porq̃ vncuerpo Redondo nopue de traer aotro tras si Si aquellos dos notienen asperas superficies y los cuerpos celestes son Re do̅dos y notienen superficies asperas Mas dizen que este mouimiento de estas speras ynferiores e̅ causado de alguna virtud que el primũ mobile

enellas ynfluye O por alguna causa natural
asi como acontece quelos cuerpos graues epessa
dos natural mente suben enalto por henchiral
guna cossa vazia Oque aquel mouimiento noes
forçoso ni natural mas milagroso ÷

✠ ℂLICENÇ Como se prueva que elcielo se
mueve de oriente en ocidente.   § COSM̄

Orque vemos quelas estrellas que
nacen en oriente poco apoco vnas
enpos deotras suben hasta que vie
nen al medio cielo y sienpre en vna misma dis
tancia e ygual propinquidad vnas con otras ya
si vniforme se van de cindiendo hasta que se nos
ponen enel ocidente Tanbien vemos quelas estrellas
que estan cerca del polo artico las quales nunca
senos asconden mas sienpre las vemos mouerse
de vna misma manera describiendo sus circulos
cerca del polo deleuante enponiente y sienpree
tan envna distancia vna conotra Delo qual parece
manifiesto assi enlas estrellas que senos ponen co
mo delas que contino vemos que el firmamēto

o octaua spera se mueve de, oriente e ocidente

LICENÇ. Pregunto si es Redondo el cielo
o que figura tiene. COSM.

El cielo es Redondo sperico lo qual
se prueua por tres Razones. La pri
mera es semejança. La segunda pro
uechosa. La tercera necessaria. La primera seme
jança porque el mundo sensible es fecho a seme
jança del mundo ar thetipo enel qual no ay pri
cipio ni fin y por esto a la semejança deste mũdo
el mundo sensible tiene figura o forma Redõda
sperica en la qual no se puede asignar o señalar
principio ni fin. La segunda que es por vtilidad
e prouecho porque en todos los cuerpos y so pe
rometros la spera es mayor cuerpo y de todas
las formas la forma Redonda es mas capaz
porque el mayor cuerpo es el Redondo siguese q
es el mas capacissimo y como el mundo todas
las cossas tenga tal forma le fue vtil e prouecho
sa. La tercera por necessidad porque si el mũ
do fuese de otra forma que la Redonda siguirse
ya que algun lugar estaria vazio o algun cu

erpo sin lugar lo qual todo es ynposible y si el
cielo fuese llano aquella parte que estouiese sobre
nuestras cabeças ser nos ya mas cercana que las o
tras yla estrella que alli estouiese ser nos ya mas
cerca quelas del oriente O ocidente

6 **LICEN.** Pregunto si tiene color el cielo O
que es la color que vemos ~   COS M

NVestros sentidos muchas vezes se en
gañan y la vista se puede mas ayna
engañar que ninguno delos otros y
assi vemos que vn palo entero puesto enel agua
parece quebrado dos torres juntas miradas de
lexos parecen ques vna El sol avnque es mu
chas vezes mayor qla tierra parece pequeños por
que es desaber que sobre laluna ninguno pue
de ver sino las estrellas yestas vemos mediante
la lumbre que tienen Recebida del sol y quando
miramos arriba como el Rayo de nuestra vista
no halla cossa firme enque afirmarse como haze
quando mira enla pared oen otra cossa semejante
desfallece y como no puede conoscer color echa el
ojo la color de su propio vmor yasi le parece q

el cielo tiene color Mas los cielos desi propios no
tienen color ninguna saluo vna pura claridad cõ
que el sol los alumbra —

7 ¶ Licenē Las estrellas q̃ tamañas sõ Epor q̃
vemos algunas corriendo cõvn Ramo de fuego : cōs.

7
Las estrellas segun dicho sea estan en
el octauo cielo yalli es su propio lugar.
y por la gran distancia que ay de nos ae
llas parecen pequeñas mas sabed que son tangra
des que si alguna cayese toda la tierra cubriria y
a quello que se veé de noche que va corriendo cõ
vn Ramo de fuego no es estrella sino fuego como
la vista lo prueua y este fuego se haze enel ayre
por a quella manera que se hazen los Relanpagos
como adelante se dira causase opor q̃ las nuves
hazen aquel fuego quando se tocan Rezia mente
delos vapores dela tierra que son dispuestos aesto
encendidos por los Rayos del sol e desfazen se luego
porque no tienen materia enque duren y estos fue
gos tanbien se hazen de dia mas por el Resplandor
del sol no parecen de dia pero si algunas vezes ti
enen tanta fuerça que su Resplandor sobre al ·

claror del sol o del dia parecense de dia y asi se lee
que ciertas vezes se vieron de dia. Estos fuegos quã
do son muchos e buelan adiuersas partes se tienen
por señales de tenpestades e tormentas de mar ~

¶ PILOTO Pues e la navegaciõ
de la mar mitamos el altura delos polos pr eqnito
q̃ son polos. § COSMOGRAPHO ~

Polos son dos puntos que ymaginamõs
en el firmamento. LLamanse polos del
mundo porque son los estremos sobre q̃
el spera del mundo su exe termina. Y asi es de notar
que entre estos dos polos ymaginamos toda la Re
dondez del cielo diuidir se en cinco zonas. La primera
es del polo artico hasta el circulo artico. La 2ª. del
circulo artico hasta el tropico de cancer La 3ª del
tropico de cancer hasta el tropico de capricornio
por medio desta passa la linea equinocial. La 4ª.
del tropico de capricornio hasta el circulo antartico
La 5ª del circulo antartico hasta el polo antartico.
e assi proporcionadamente la tierra se diuide en cin
co partes. a estas cinco zonas supuestas Y es de sa

53

.ber que porque la Redondez del cielo contiene
trezientos y sesenta grados dezimos que de vn
polo a otro ay ciento yochenta grados y notad
quelos que abitan debaxo dela equinocial por
que estan en medio del mundo y ygual mente tie
nen entranbos polos por orizonte. Mas los que a
bitan fuera dela equinocial ala parte del vn po
lo odel otro no ve en mas del vn polo yasi nos
vemos enel cielo el lugar do el vn polo ymagina
mos el otro senos esconde. Pues este que vemos
se llama por estos nonbres polo artico / Septen
trional boreal. LLamase polo artico de artos
que es mayor osa. Septen trional sellama delas
siete estrellas que estan cerca deste polo. E dizese
boreal por vn viento llamado boreas que desta
parte nos viene al qual llamamos norte. El otro
polo contrario se llama Antartico. Austral. e
Meridional llamase antartico de ante que es cotra
casi puesto encontra del artico. Austral por que del
nos viene vn viento que llamamos austro. Ella
mase meridional porq esta ala parte demedio dia
quanto anuestra abitacion ~

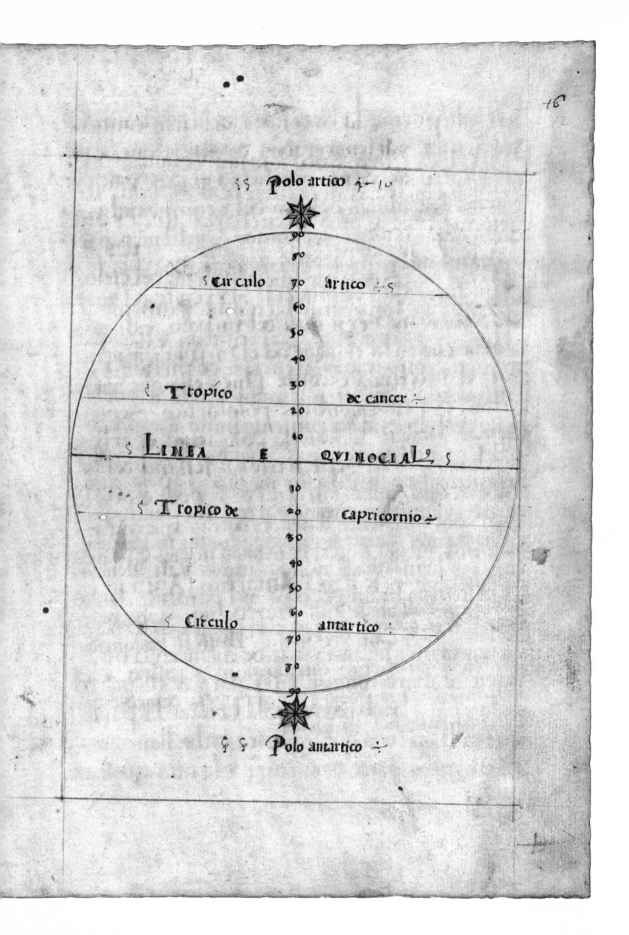

Polo artico

90
80
Circulo   70   artico
60
50
40
Tropico   30   de cancer
20
10
LINEA   E   QVINOCIAL
10
Tropico de   20   capricornio
30
40
50
Circulo   60   antartico
70
80
90

Polo antartico

¶ PILOTO. Porq̃ q̃ndo tomamos el altura del norte llamamos ala cabeça parte de encima y alpie parte de abaxo. Como ay enel cielo cabeça y pie

COSM

Es de saber quelas partes del cielo son quatro Leuante Poniente Medio dia Septentrion. Estas quatro partes conparadas al cuerpo de vn honbre llamamos ala del leuante son q̃ parte mas noble braço yzq̃erdo y ala de poniente braço derecho y al septentrion cabeça y al medio dia pie. Ey maginada vna Raya por medio del polo que diuida el mundo endos partes ala parte que es hazia nos llamamos cabeça ala otra parte pie ça y por la misma Raya los braços y asi viene a ser el leuante braço ysq̃erdo yel poniente braço derecho. Pues destas dos partes q̃ esta Raya hazen la vna sellama parte de encima y la otra parte de abaxo (conviene saber) parte de encima del polo y parte de abaxo del polo y destas la parte que es hazia nos (que segun sea dicho llamamos cabeça xzimos parte de encima y la otra que sea di

cho ser y pie dezimos parte de abaxo La Razõ
es porque los que abitan debaxo dela equino
cial tienen los polos por orizonte segun dicho
es y como todas las estrellas tanbien las que estã
en medio cielo como las que estan cerca delos
polos cada dia les nacen y seles ponen la estre
lla del norte dando su buelta Redonda al po
lo quando haze la media buelta dende la Raya
que dicha es por esta nuestra parte que es cabeça
dezimos ser encima del polo porque entõces
aestos que abitan enel medio del mundo les es
ta encima del polo ela podrian ver y quando
haze la otra media buelta que es ala parte
del pie esta debaxo del polo porque en
tonces nola podrian ver y así sea de entender
loque dezimos aver enel cielo cabeça y pie
y ser la cabeça parte de encima el pie parte de
abaxo Por lo qual tomando el altura del
norte quando el estrella esta debaxo del po
lo a crecentamos aquellos grados y quantos
esta encima quitamos dela misma altura ÷
la figura desto parece enla plana siguiente ÷

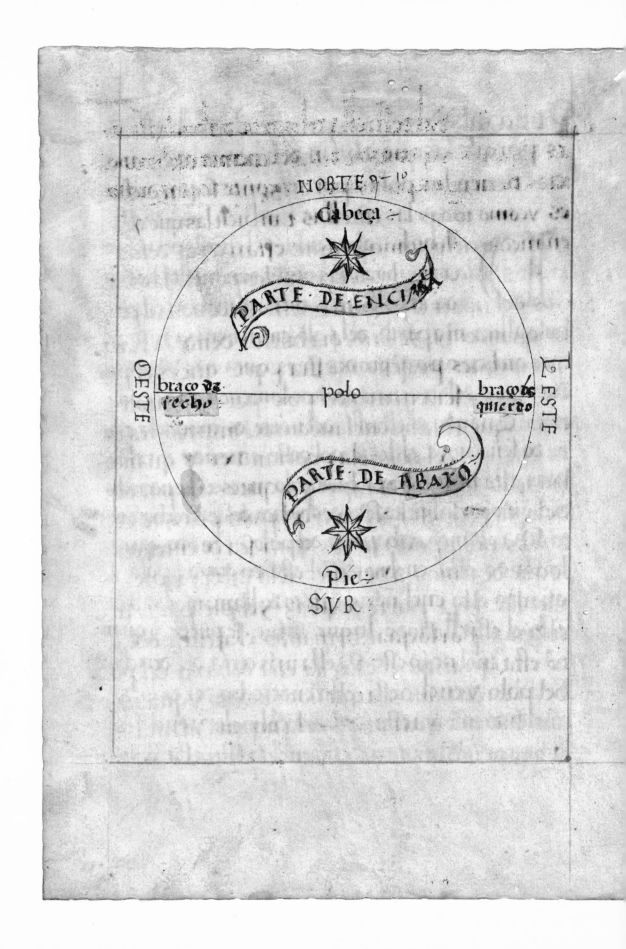

NORTE

Cabeça

PARTE · DE ENCIMA

OESTE   braço derecho   polo   braço izquierdo   LESTE

PARTE · DE · ABAXO

Pie

SVR

58

¶ PILOTO Si la estrella del norte da buelta Redon
da al polo porq̃ vnas vezes esta encima o debaxo del
mas de tres grados yotras menos de medio grado

COS M

Dicho sea que tan bien el estrella del norte
como todas las otras dan buelta Redon
da al polo mas enesta buelta el estrella
nose allega ni aparta del polo en vn tienpo ni lugar
mas que en otro Pero dezimos que vnas vezes esta
encima o debaxo mas grados o menos enesta ma
nera Quando el estrella del norte es en aquella par
te do señalamos el leste o el oeste entonces no esta
mas alta ni mas baxa quel polo antes esta enderecho
del y quanto por su buelta se va desbiando deste derecho tan
to se ba alçando o abajando del polo de modo que
quando esta enel noroeste esta encima del polo y
quando esta enel norte que es do llamamos cabeça
esta el estrella todo lo que subir se puede y quan
do esta enel noro este ya esta mas cerca del derecho
del polo y enel oeste esta enderecho del polo y
enel sudueste ya esta debaxo del polo y enel sur
lo que mas abaxar puede y enel sueste esta debaxo

ura del polo De modo que quando la estrella es
touiere enlos dos Rumbos deleuante oponien
te esta nimas alta nimas baxa que el polo y
enel noroeste norte enoroeste esta encima
del polo y enel sudoeste sur esueste esta de
baxo del polo y asi se entiende estar lacstrella
mas grados omenos debaxo oencima del polo
y en quanto al quitar oponerse estos grados
enel altura notarsea que quando tomamos
el altura del norte es para saber quantos gra
dos esta leuantado el polo sobre nuestro orizon
te y esta altura tomase al estrella y no al polo
porque el polo nolo vemos yla estrella como
haze labuelta que dicha es la media buelta es
ta encima del polo yla otra media debajo Asi
quando esta debajo, lo que le falta para estar
enderecho del polo leacrecentamos sobre los
grados dealtura que tomamos y quando esta
encima aquellos grados que esta mas alta q
el polo quitamos dela misma altura que a
vemos tomado ꝉ ꝉ ꝉ exenplo s.

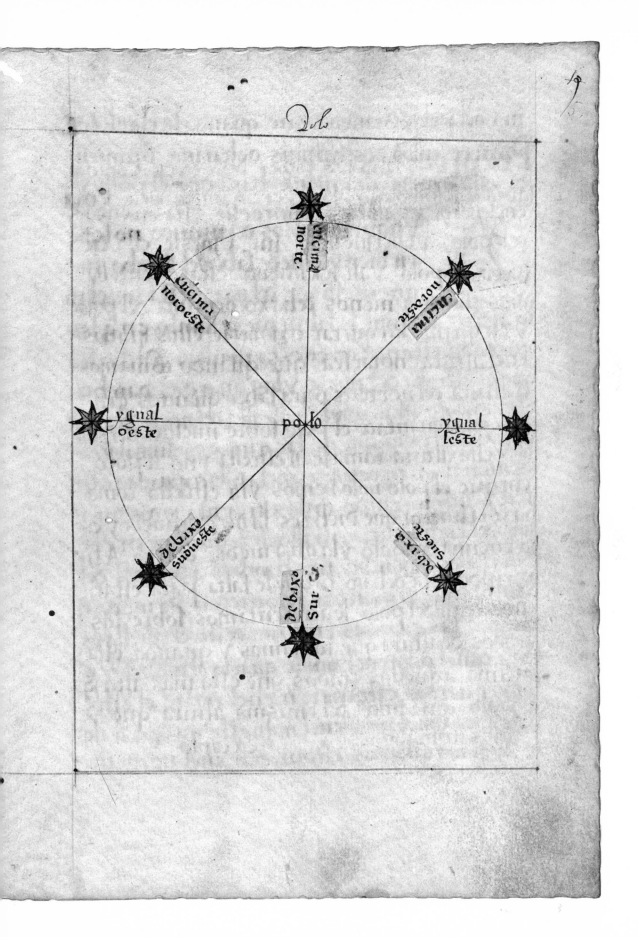

¶ PILOTO Pues los polos nose veen Como se
sabra el Runbo enq̃ esta coel el estrella del norte. COSM.

Unquelos polos del mundo nose ve
en bien se puede saber el Runbo en
que el estrella del norte esta conel po
lo artico lo qual se sabra por el Runbo en q̃ las
guardas estouieren enesta manera. Dicho sea
que el estrella del norte ylas guardas dan bu
elta alpolo Mas como el estrella del norte sea mas
cercana al polo que no las guardas su buelta
es menor en Redondez quela q̃ las guardas ha
zen Pues mirando enlos treynta ydos vien
tos dela nauegacion hallarsea que sienpre el
estrella del norte va tres Runbos a delante del
Runbo enq̃ las guardas estouieren. De modo q̃
estando las guardas enel nordeste el estrella es
tara conel polo enel sur quarta al Sudueste E
si las guardas estouieren enel norte el estre
lla estara conel polo enel nofueste quarta al noi
te. Y si las guardas estouieren enel noroeste

quarta al oeste la estrella estara enel oeste ni
mas alta ni mas baxa quel polo y por estos
Rumbos se podra saber enlos demas el Rumbo
enque el estrella del norte esta conel polo

Polo

guarda

Polo

9 10

¶ PILOTO Pues por las guardas se conosce q̃ o
ra es dela noche. Si el norte esta tã bajo q̃ las guardas
(ẽ parte) no se vean como se sabra q̃ ora es? · COS M̃·

Es de notar que para tener Conosci
miento y saber Justa mente que
ora es dela noche demas delas gu
ardas ay otras tres estrellas por las quales a
vnque las guardas no se vean por estas o por
qual quier dellas se sabra que ora es asi como
por las guardas se sabe En esta manera. Estas
estrellas se llaman Tercera Sesta Nouena E
tienen estos nonbres por que la tercera anda
tres oras atras dela guarda y la sesta seis y la
nouena nueue. De modo que estando la guar
da en la Cabeça la estrella tercera estara en el nor
deste y la sesta en el leste y la nouena en el suest
y por estos Rumbos se sabran los demas y asi
conoscidas estas estrellas aunq̃ las guardas

no se vean por estas se conoscera la ora que es ar
bitrando el Rumbo en q̃ cada vna va y sabido asi
mismo donde la guarda a de hazer la media noche
y el conocimiento destas estrellas servira asi mismo
pa tomar el altura del polo q̃ndo el estrella del norte
estoviera tanbaxa que no se pueda ver la buelta
delas guardas Ca por estas se sabra donde la guarda
esta ÷

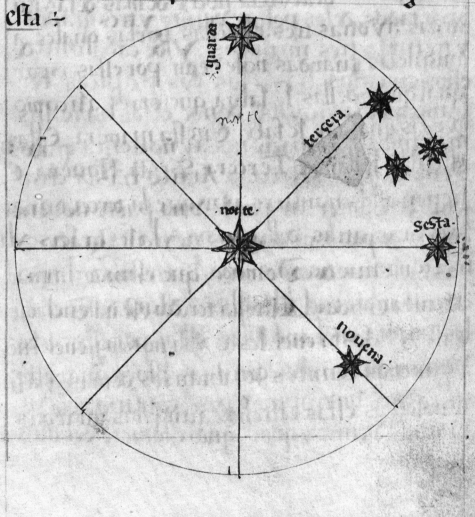

¶ **PiLoto**. Que cossas son Circulos Artico
y antartico ~ · COSM·

Circulos son dos puntos y magi
nados vno en circunferencia del
polo artico yotro encircunfere
cia del polo Antartico Y estos Circulos estan
apartados delos polos veinte y tres grados
y treinta y tres minutos. Y de cada vno de
estos circulos hasta los dos tropicos de
Cancer y capricornio donde el sol haze sus
maximas declinaciones ay nouenta grados
De modo que del circulo Artico hasta el tro
pico de Cancer ay por la vna parte nobenta
grados y por la otra otros nouenta grados
hasta el tpico de capricornio lo mismo dezimos del
circulo antartico los q abita abaxo destos circu
los vna vez enel año veen el sol por espacio de
veinte y quatro oras cotinuas esto es quado esto
uiere enel vn tropico y por el cotrario nolo ve
ra otras veinte y quat quado estouiere e el otro
tropico ~

22

14 ¶ PILOTO Pues los tropicos estan apar
tados el vno del otro quarenta ysiete grados
y seis minutos Como dende qualquier circulo
hasta entranbos tropicos ay nouenta grados
justos ~

14

§ COSM̃ ·

14

E S de notar que en qualquier lu
gar quel honbre este vee la mi
tad del cielo ylamitad sele asco̅
de y como el cielo tenga trezientos yseseñ
ta grados siguese que vee ciento yochenta gra
dos enesta manera los nouenta delante desi
yotros nouenta tiene tras desi yacada lado
nouenta de modo que el esta en medio delos
ciento yochenta yasi de cada parte que mi
rare vera nouenta grados del cielo que llegan
asu orizonte. Pues estando el honbre enel
circulo artico digo q̃ vera entranbos tropi
cos a cada nouenta grados. Enesta manera
Mirando ala parte del norte vera el tropico

de cancer en nouenta grados que es su orizonte los
veinte y tres grados y treinta y tres minutos q̃
ay desi hasta el polo y sesenta y seis grados y ve
inte y siete minutos que ay dende el polo al tro
pico que son nouenta grados asi que su orizõ
te por aquella parte es justa mente por el dicho
tropico de Cancer y buelta la cara al sur vera
el tropico de capricornio en otros nouenta gra
dos los sesenta y seis grados y veinte y siete mi
nutos que ay del circulo ala linea equinocial
y veinte y tres grados y treinta y tres minutos
que ay dela linea al tropico q̃ suman nouenta
grados Asique des de el mismo circulo artico
Su orizonte llega al tropico de cancer y al tropi
co de capricornio De modo que en qual quier
delos circulos que honbre este vera entrabos
tropicos a nouenta grados Justamente =

¶ PiLoTo. Que cosa es orizõte.   COSM
Rizonte es vn circulo grande en derre
dor del mundo enla super ficie o sobre

haz dela tierra llamase orizonte que quiere de
zir terminador de nra vista · por q̃ del nuestra
vista termina e vee la mitad del cielo ques so
bre este nuestro hemisperio dela otra mitad
que es sobre el hemisperio baxo ~

¶ Piloto . Que cossa es linea equinocial E
por q̃ se busca enel altura   ¶ COSM

Linea equinocial es vna Raya y
maginada por el medio mundo y
gual mente apartada delos polos
E dizese equinocial por que passa el sol por e
lla dos vezes enel año es asaber a onze de março
y treze de setienbre e haze equinocio que quie
re dezir ygualador esto es que entodas partes los
dias y noches son yguales · y de alli se ha llegan
a vno delos tropicos y es de saber quelos que
moran debaxo dela Redondez desta equinocial
tienen spera derecha por que ygual mente se
les alça los polos · y tanbien se dize derecha por
que su orizonte parte la equinocial en angu
los derechos Y aquellos que moran dende la e

69

quinocial aesta parte nr̃a quellamamos par
te del norte O dela otra quellamamos parte
del sur dezimos tener spera torcida porq̃ el
vn polo se alça y el otro se abaxa · y por que
el orizonte Corta la equinocial en angulos
yn yguales · Esta linea equinocial se busca ẽ
la nauegacion delamar conel altura del sol y
del norte paraque sabidos los grados que el
que nauega esta desviado della alaparte del nor
te oala del sur Asi toma la derrota conforme a su
camino ÷

¶ PiLoto Que son los tropicos E porq̃ se di
zen assi   ¶ COS M̃

Ropicos son dos puntos oluga
res donde el sol vltima mente lle
ga vno ala parte del norte yotro
ala parte del sur Enesta manera entrando el
sol enel primer punto del sino de Cancer por
arrebatamiento del firmamento describe o
causa vn circulo el qual es el que vltima mẽ
te haze dela parte del polo artico Ellamase
circulo de estiual solisticio Ollamase tanbien

tropico estiual y en eneste punto es lo q̃ el
sol esta mas cercano anos . Tanbien el sol
entrando enel primer punto del signo de
capricornio discribe vn circulo que vlti
ma mente haze ala parte del polo antartico
o parte del sur el qual sellama Circulo del
so listicio yemal . y el sol entonces es apar
tado de nos   Cada vno destos tropicos esta
desviado dela linea equinocial veinte y
tres grados y treinta y tres minutos Asi q̃
ay del vno alotro quarenta y siete grados
y seis minutos esta es la latitud dela terce
ra zona ala qual los antiguos llamaron to
rida que quiere dezir tostada por q̃ sienp
el sol haze su mouimiento por ella dentro
destos tropicos . E tropico se dize de tropos
en griego que quiere dezir conbersion por
que andando el sol por su zodiaco en llegan
do alos tropicos se convierte e buelue a tras

¹⁸ **PILOTO**. Que es zodiaco   COSM

¹⁸ **Z** O diaco es vn circulo enel firma

mento que corta la equinocial ytanbien es
cortado della endos partes y guales la vna
parte declina hazia septentrion yla otra
hazia el austro Partese endoze partes yqui
ales. E cada parte destas sellama signo. E tie
ne nonbre especial de animal ode otra cossa
conquien conviene ensemejança o propiedad
Y estos doze signos del zodiaco se Refieren a
la naturaleza del sol Ca segun el efecto que
haze quando esta encada parte deaquellas.
tal nonbre pusieron alsigno. Cada vno destos
signos tiene treinta grados delongitud ydoze
grados delatitud. Donde parece quel zodiaco
tiene enlongitud trezientos ysesenta grados
Este zodiaco se entienda vna sobre haz ÷ tie
ne los dichos 3. 6 o grados encircunferencia
ydoze enanchura La parte del zodiaco que
se aparta dela equinocial hazia el septentriõ
sellama septentrional ylos seis signos que
son dende principio de aries hasta el fin de
virgo sellaman septentrionales La otra parte
del zodiaco quese aparta de la equinocial ha

zia el austro sellama meridional yaquellos se
is signos queson dende principio delibra has
ta el fin de piscis sellaman meridionales o a
ustrales.

¶ LICENCIADO

19 ¶ Que cossa es el sol E porq̄ se dize asi E si solo
su mouimiento fuera mas prouechoso q̄ el
mouimiento Rapto. E que sienpre andouie
ra por la linea equinocial porq̄ haze el tpō
mas tenplado

¶ COSMOGRAPHO

19

El sol es fuente delunbre. E dizese sol
porque el solo da lunbre a todas las
cossas animadas E porel tienen bida
el qual si solo semouiera por su propio moui
miento ques de ocidente en oriente en bn año
hiziera vna buelta al mundo de modo quese
is meses continuos estouiera sobre nuestro
hemisperio que no touieramos noche y otros
seis meses estouiera en el hemisperio baxo q̄

73

no tomieramos oia Porlo qual fue muy con
veniente que para caufar las oiferencias oe
los tienpos oelaño fe hiziefe el mouimiento oel
fol por toda la Reoonoez enbeinte y quatro o
ras y fiel fol anoouieta fienpre por la equi
nocial oe mas oe otros oaños q̃ fe caufara los
frutos oela tierra fin los quales no pooemos bi
uir no nos maouraran y no tubieramos los
bienes quel fol caufa enhazer nos como haze
los oiferentes tienpos que tenemos enel año ⸗

20 ¶ **Li cenc̃** Como caufa el fol los tienpos
20 oel año ⸗ **COS M̃**

CAufa nos el fol los tienpos oel año
allegandofe y apartandofe oe nos
enefta manera Quando elfol es
mas lexos oe nofotros no nos comunica tanto
fu calor como quando es cerca por loqual como
la tierra y el agua natural mente fean frias la
tierra enfriafe mucho porque no ay calor conq̃

el ayre se adelgaze y el ayre conel frio espesase e ha
zese niublados y quando se deshazen tornanse e
luvia y assi se haze la parte del año que llamamos
ynbierno que es frio e vmedo y eneste tienpo a
prietase la tierra por el frio y su calor acidental
queda dentro y cria las Raizes delos arboles
Y quando el sol es venido hasta aries que es en
março como noes muy lejos ni muy cerca de nos
asi el tienpo nonos es muy caliente ni muy frio
mas es tenplado Casi haze la parte del año que
llamamos prima vera y eneste tienpo los/ agujse
ros dela tierra q eran cerrados porel frio del yn
vierno abrense conel calor del sol e passa el calor
alas Raizes y atrae el vmor y deaqui viene que
vn mes deste tienpo llamamos abril que quie
re dezir abriente porq la tierra seabre enla mane
ra que dicha es y es propiedad deste tienpo no
estar firme Mas agora lluvioso y luego claro
y asi agora caliente y luego frio Y quando elsol
es venido a cancer que es enjunio como es mas
cerca de nos que en otro tienpo seca la tierra

Y causa la parte del año que llamamos estio ḡ
es Caliente y seco y la natura deste tienpo es se
car las yervas e las hojas delos arboles Casi
son causados por el sol estos tienpos del año.

21 ℂ LICENC̄. Si la diversidad delos tp̄os es por
estar el sol cerca o lexos de nos. Quado es en leo tāto
esta de nos como en giminis y en virgo como en ta
urus. y en libra como en aries Pues porḡ estando en
geminis taurus e aries el tpo haze tenplado. y en leo.
virgo e libra muy caluroso.   ·COS M

21 YA sea declarado que en la prima ve
ra que anda el sol en los tres signos
de Aries taurus geminis ḡ son en
março abril e mayo En estos por estar cerca del
tienpo caluroso que viene y por el frio del yn
vierno que es passado. es el tienpo tenplado—
Mas quando el sol anda en leo e virgo que es en
Julio e agosto es calienta nos mucho por que
la humedad del ynvierno ya es gastada por el ca
lor del estio y por esto este tienpo es tan calien
te y seco y quando el sol deciende a libra que es

en setienbre la Calor es quasi muerta y asi
causa el tienpo frio y seco ~

22 ¶ LICENC Porq̃ vn estio O ynvierno es mas frio
o mas seco que otro O mas vmedo. ¶ COSM.

22 Abed que asi como el sol haze los
quatro tienpos del año que dicho e
tanbien los otros planetas hazen
sus estios y sus ynviernos De donde quando el
sol nos haze estio y se sigue que algun planeta ha
ze su ynvierno el estio no estan caliente ni tan
seco y si otro planeta junta mente conel sol ha
ze su estio es mas Caliente y mas seco que en otro
año Por lo semejante quando el sol nos haze in
vierno si otro planeta haze su estio el ynbier no
haze menos frio e menos vmedo y si el otro pla
neta haze su ynbierno conel sol el ynbierno es
mas frio e mas vmedo E por esta Razon se juz
gue delos otros tienpos Tanbien sabido es q̃
ay planetas Calientes/e frios/y secos/e vmedos
y por esto si enel ynbierno ay conel sol planeta q̃
sea Caliente y seca haze el ynbierno menos frio
e menos vmedo E si enel verano ay conel sol

planeta frio evmedo el verano es menos Caliente y
menos seco·

23 ¶LICENCP orq nosparece el sol mayor en
oriente O eno ccidente que no enel medio cielo
pues enel medio es mas cerca denos q̃ enlas
otras partes· COS M

A S ies quel medio cielo es mas cerca
denos quelas otras partes· por q̃ entre
nos yel medio cielo ay solamente aire
y fuego yentre nos yla parte oriental E ocidẽ
tal ay estos dos elementos y mas lamitad dela ti
erra De modo q̃el medio cielo es mas cerca de nos
quelas otras partes· Mas lacausa desta apare
cia del sol es por los vapores que suben entre
nuestra vista yel sol y como aquellos vapores
sean cuerpos diaphanos desgregan los Rayos
de nuestra vista detal modo que no pueden con
prehender lacossa ensu propia cantidad· asi co
mo parece dela moneda echada enel agua cla
ra· que por ladesgregacion delos Rayos no pa
rece del tamaño que es y que el sol este mas

apartado denos envna parte del cielo que en
otra nocontradize lo que de suso sea declarado
ser el cielo Redondo porque el cielo Redondo es
ygual mente apartado del centro del mundo e
nosotros somos enla superficie osobre haz dlatra.

24 ¶ LICENÇ Leese que muchos soles sevieron
juntos E lunas. Pregunto si esto pudoser osi
se engaño lavista.

COSM

D Exado lo que por milagro sabemos q
fue hecho. Mas natural mente digo
quesi diuersos estanques deagua.
son cerca denos encada vno vemos la figura
del sol o dela luna. Pues biena si podemos ver
la figura del sol o dela luna. envna nube que no
es otro sino agua y a questa ymagen llaman
los los griegos parelion. y tiene semejança ala   parelios.
figura del sol ensola la grandeza pero no lunbre
ni calor y asi sepueden hazer tantas figuras
quantas nuves oviere oppuestas alsol oala luna
es dispusicon para Recebir aqlla semejaça

¶ PILOTO.

En la navegacion dela mar nos Regimos por
el altura del sol y esta altura se toma cōforme
a las sonbras q̃ haze Pregunto quantas diferen
cias de sonbras haze el sol enel mūdo.

¶ COSMOGRAPHO

L As sonbras principales que el sol ha
ze alosque abitan enel mundo son
çinco Es a saber sonbra al leuante
Sonbra al poniente Sonbra al norte sonbra al
sur e sonbra derecha Dezimos sonbra al leuāte
yes quando el sol es enel poniente que nuestra
sonbra entonce va al leuante y dezimos sonbra
al poniente quando el sol nos es al leuante q̃
nos nace Ca entonces nuestra sonbra va al po
niente Dezimos tanbien sonbra al norte yes
quando el sol nos esta al medio dia q̃ entonces
nuestra sonbra derecha mente va al norte La sō
bra al sur es alos que abitan ala parte del sur
quando el sol anda mas al norte Ay tanbien

sonbra derecha y es quando el sol esta sobre
nuestra Cabeça Destas sonbras los que
abitan dentro delos tropicos tienen enel
año todas cinco sonbras Es asaber al leuãte
quando el sol seles pone y al poniente q̃ndo
el sol les nace E asi mesmo al norte y al sur
y sonbra derecha porque el sol dos vezes ẽ
el año passa por cima desus cabeças y entõ
ces su sonbra no dechna aparte alguna Los
que abitan debaxo delos tropicos tienen qua
tro sonbras Al leuante y al poniente y los
del tropico de Cancer sonbra al norte y los
del tropico de capricornio sonbra al sur y una
vez enel año sonbra derecha quando el sol
esta encada vno delos tropicos Los que
abitamos fuera delos tropicos tenemos tres
sonbras es asaber al leuante y al poniente
y los q̃ estamos ala parte del norte sonbra
al norte y los dela parte del sur sonbra al sur
mas nunca tenemos sonbra dcha

26 ¶ **PILOTO** Pregunto si ay algun lugar
donde avnque se vea el sol no haze sonbra
26 a ninguna destas partes ~ ⸿ COS~

Igo que ay parte donde avnque
se vee el sol no haze sonbra al leuā
te ni al poniente ni al norte ni al
sur ni sonbra derecha. lo qual se prueba asi
Si alguno estouiese muy precissa mente
debaxo del polo artico. este ni ternia parte de
leuante ni parte de poniente. por que el sol
no se le leuanta de donde viene el nonbre de
leuante ni se le pone de donde viene el nõbre
de poniente. y no teniendo estas dos partes
no tiene parte de norte. antes põtual mēte
esta debaxo del ni parte de sur por que lo tiene
so los pies. Pues notorio es que este veria el
sol seis meses del año Mas avnque viese el
sol no ternia sonbra aninguna delas partes
sobre dichas. ni sonbra derecha por que nunca

el sol en ningun tienpo puede passar por su cenic

27 ¶P̃tlolo Si puedo estar enalgun lugar tā
aparta do del sol como del norte~   ·COSM·

27

Ien puede estar tan apartado
del sol como del norte encierto
tienpo ylugar enesta manera
Siestais enseuilla alos veinte yvno deotu
bre estareis tan apartado del sol como del
norte porque seuilla dista dela equinoci
al treinta yocho grados ala parte del polo
artico ydese mismo polo cinquenta ydos
y el sol este dia esta apartado dela equinoci
al Catorze grados a la parte del polo antar
tico de modo que de seuilla ala linea treinta
yocho y del sol ala linea Catorze son cinquē
ta ydos grados Asi que eneste lugar ytienpo
esta tan apartado el sol como el norte ¶ por
este modo se puede entender deotras partes
conformando el lugar conla dedinacion
q̃ el sol haze asi ala parte del norte como delsur ·/·

¶ LICENCIADO

28 ¶ Como se causa el eclipsi del sol E porq̃ mas
en vn tienpo que en otro ⁖

¶ COSMOGRAPHO

28

Vando la luna esta enla cabeca
o cola del dragon o cerca si esto ui
ere en conjuncion conel sol entõ
ces el cuerpo lunar se ynter pone entre nr̃a
vista y el cuerpo del sol y como la luna sea tene
brosa desi mesma· ynter puesta entre el sol y
nos quitanos la claridad del sol yasi el sol
p̃adece eclipsi· no porque el carezca delunbre
mas nosotros carecemos de su lunbre por la
ynter pussicion dela luna entre nuestro aspe
cto y el sol· el qual eclipsi sienpre se causa ẽ
la luna nueba· delo qual parece manifiesto
que como enla passion del señor el eclipsi que
hizo el sol fue enla luna llena· aquel eclipsi
nofue natural mas milagroso ccontra natu
ra por que como es dicho en el tienpo q̃la luna
es nueba y no en otro ade acontecer ⁖

29 ¶ **LICENÇ** Pues muchas vezes hablamos del
29 tienpo· Pregunto q̃ cossa es tienpo~   **COSM̃**

Tienpo es la tardança del mouimien
to delos cuerpos celestiales El q̃l
durara tanto quanto durare el
mouimiento delos ciclos De donde se ynfiere
que despues del Juizio final como cessare el
mouimiento celestial no abra mas tienpo ni
abra mas alguna diferencia de tienpo y es de
notar q̃ el tienpo no fue hecho en tienpo·
porque a hazerse en tienpo era menester q̃ el
tienpo en que se hizo fuese hecho en otro tie
po y aquel en otro de modo que seria proce
sso ynfinito e Es de saber q̃ el tienpo nōbra
mos en cinco partes principales que son :
Año· Mes· semana· Dia· Ora·

30 ¶ **LICENÇ** Que cossa es año E por q̃ se di
30 ze assi~   **COSM̃**

Ano es todo el tienpo q̃ el sol pas
sa los doze signos del 30 dia co y bu

duc al punto do començo lo qual haze en
trezientos y sesenta y cinco dias y seis oras E
dizese año quasi anulo que es lo mismo que
circulo porque buelve a lo mismo que come
ço Los egipcios antes que tuviesen vso de
letras acostunbrauan figurar el año en vn
dragon que mordia la cola pero despues que
començaron a tener año començaronlo en
setienbre porque entonces los arboles tie
nen fruto Lo mismo hazen los arabes L
Los hebreos Comiençan el año en março
porque les fue dado por ley al qual llaman
año legitimo Nosotros lo començamos
en enero porque entonce el sol comiença a
boluerse a nos

31 ¶LICENÇ Mes que cossa es E de do to
ma nonbre ÷   COSM

31

Mes es mesura o medida que mi
de el año y viene de mene que

es nonbre griego que significa luna. y de
a quies que se llama mes todo aquel espa
cio de tienpo que la luna apartandose del
sol se viene ajuntar conel acabado su circu
lo. A los meses los Romanos les pusieron
nonbres. delos dioses que ontrauan. laql
denominacion nosotros seguimos. Los he
breos contavan el mes por la luna dizie
do el primero dia de luna primero del mes
y asi quantos dias eran de luna. tantos.
tenian del mes. Enlas otras naciones tie
nen los meses. nonbres. particulares. Edi
ferentes.

CLICENĊ Que cossa es semana E dedo
se dize assi   COSM

Emana es vn numero de siete
casi contiene siete dias natu
rales enque dios hizo todas
las cossas enlos seis y folgo enel septimo y
asi quiso que fuese sanctificado. Estos.

siete dias dela semana toman nonbre delos
nonbres delos planetas. y por que el sol Rey
na enla primera ora del domingo Sellama do
minica. y por que la luna enla primera ora
del lunes Sellama de su nonbre. yassi delos
otros. La iglesia Catolica acostunbra con
tar estos dias dela semana por ferias assi
como segunda feria por lunes. tercia feria
por martes. yasi delos otros dias ecepto sa
bado y domingo. Los hebreos los nòbra
uan por nonbres numerales. asi como pri
ma sabati por domingo segunda sabati por
lunes. e por consiguiente de todos los otros.

33 ¶Licenc̄ Que cossa es dia E quantas ma
33 neras ay de dia ÷   § COS M.

Dia es lalunbre o claridad que
elsol causa enel mundo E dizese
dia adijs que sirnifica los dioses
de quien los dias tomaron nonbres o lla

88

manse deste nonbre dia · engriego que en
latin significa duo · por que el dia es conpues
to de noche eluz · O dizese a dian que signi
fica claridad olunbre · y es de saber que dia
se entiende endos maneras · q̃ son dia natu
ral / que contiene veinte y quatro otas · yẽ
estas se yncluye noche ydia · E dia artificial
que es el tienpo quel sol esta en nuestro he
misperio · El dia natural tiene quatro calida
des que son estas dela nouena parte dela no
che hasta la tercia parte del dia Caliente y
vmedo y dela tercia parte del dia hasta la ·
nona Caliente yseco · y dela nona hasta la
tercia parte dela noche frio yseco ydela ter
cia parte dela noche hasta lanouena frio e
vmedo Assi mesmo el dia artificial que el
sol nos alumbra tiene quatro partes La
primera parte del dia el sol parece bermejo ·
Enla segunda Resplandece · Enla tercera es
calienta · Enla quarta deciende e atibiase E

por esto dauan al sol quatro cauallos segũ
estas quatro diferencias que haze. Los egip
cios Comiençan el dia quando se pone el sol has
ta otro dia a la misma ora. Los griegos e per
sas lo comiençan dela mañana Los Romanõs
dela media noche. hasta otro dia al mismo
punto Los astrologos atenienses carabes
del medio dia. La yglesia catholica para ce
lebrar las fiestas toma el principio delas
visperas E para la abstinencia e calidad de
los manjares de media noche a media noche
y lo mismo se entiende dela obseruacion y sole
nizacion delas fiestas. quanto a la cessacion
delos officios serbiles ÷

34 ¶ LICENÇ͂   Ora que cossa es COSM̃
   34
ra es vn espacio de tienpo en que
passa el sol medio signo del zodiaco
E porque en el zodiaco ay doze signos
y el sol passa todos doze en vn dia natural de a
qui es que en el dia natural ay veinte y quatro

90

oras E porque enla equinocial entra el
sol ygual mente enlos signos del zodiaco vi
ene enaquel tienpo a aver doze oras enel dia
e doze enla noche · ygual mente ÷

35 EL ICENC Por que vnos dias son gran
des yotros pequeños Que es la causa q̃ crece
y menguan ÷   COSM

35 Y A sea dicho que el dia artificial
se causa por la lunbre del sol · y qñ
to el sol mas se allega a nos · tanto
el dia nos es mayor e la noche menor E por el
contrario quanto el sol se nos aparta es me
nor el dia e mayor la noche · Y es de saber que
a quellos que abitan debaxo dela equinocial
que como es dicho el sol entra ygual mente e
los signos del zodiaco · estos tienen los dias e
noches yguales · por que su orizonte sienpre des
cubre los seis signos del zodiaco y estos tiene
dos veranos · por q̃ el sol dos vezes enel año passa
por el cenit desus cabeças · lo qual es enlos pñ
cipios de aries y libra · Tanbien tiene dos ·

inviernos q̃ son quando el sol esta en los tro
picos de Cancer y capricornio por que en
tonces es loque el sol mas se aparta dellos ·—
Mas los que abitan ala parte del norte cómo
el sol comiença a subir dende el primer punto
de capricornio para Cancer quanto va su
biendo tanto los dias nos van creciendo E a
los dela parte del polo antartico menguando
y llegado a aries a onze de março describe la e
quinocial haziendo equinocio · esto es en todas
partes los dias e noches yguales/y passado el p̃
mer punto de aries nos comiençan a ser los
dias mayores que las noches yala otra parte
menores y llegado el sol al tropico de Cancer
alos onze de junio hazenos el mayor dia yla
menor noche yala otra parte por el contrario
por q̃ entonce esta mas allegado a nos el sol · y de
alli torna a descendir p̃ra capricornio e como va
descendiendo esto es apartandose denos van nos
menguando los dias y creciendo las noches y

92

llegado ahora alos treze de setiembre torna
a discribir la equinocial haziendo los dias e
noches yguales / y dealli torna a decendir para
capricornio eban siendo mayores las noches q̃
los dias y como llega altropico atreze de dizie
bre nos haze la mayor noche y menor dia ya
los dela parte del polo antartico porel contra
rio porqalli el sol es loque mas se aparta denos.

36 ¶ **Lacenc** . Si por llegar senos el sol tene
mos mayor eldia Porq̃ quando el sol anda des
ta parte del norte los q̃ abita mas cerca del polo q̃
estan mas apartados del sol tienẽ mayor dia COSM.

36 Esaber es que //los q̃// abitan dela equi
nocial alos polos quanto el polo
mas seles leuanta sobrẽ su orizon
te tanto mayores les son los dias enesta ma
nera Aquellos cuyo cenic es enel circulo ar
tico alos quales el polo se leuanta sobre su o
rizonte se senta y seis grados y ueinte y siete mi

nutos· quando el sol estouiere enel primer
punto de cancer sera aellos vndia deveinte
y quatro oras y quasi vn instante por noche
porque envn momento toca el sol el orizonte
dellos y luego sale· yaquel tocamiento tiene
por noche· y al contrario acaesce quando elsol
esta enel primer punto de capricornio porq̃
entonces es aellos vna noche deveinte y qua
tro oras y quasi vn momento por dia· por que
envn instante toca el sol suorizonte yluego
seasconde yaquel tocamiento tienen por dia·
yasi porel contrario los q̃ abitan debaxo del
circulo antartico· yaquellos cuyo cenic es e
tre el circulo yel polo del mundo mientra elsol
estouiere enaquella parte que el orizonte dellos
corta del zodiaco ser les ya vndia continuo sin
noche y si aquella fuere decantidad devn sino
sera alli vndia continuo devn mes· e si fuere
de dos sinos sera vn dia de dos meses easi los
demas y los que abitasen debaxo delos polos·

todo el año les seria vn dia cvna noche. Ental
modo que el que abitase debaxo del polo ar
tico enlos seis meses q̃ el sol anda dende pri n
cipio de aries fasta en fin de virgo q̃ son los seis
signos septentrionales les seria vn dia sin noche
y por el Contrario quando el sol anda dende pn
cipio de libra hasta en fin de piscis que son los
otros seis signos australes le sera vna noche có
tinua sin dia. Assi q̃ la mitad del año le seria vn dia
cla mitad vna noche/ lo mismo se entienda del q̃
abitase debaxo del polo antartico. Ela causa
es por q̃ como la Redondez del mundo quanto
se va llegando a los polos va siendo menor. assi el
orizonte de aquellos que mas se llegan alos polos
descubre mas dela buelta que el sol da enel cielo
quando anda de aquella parte. Demodo q̃ la tic
rra no les ocupa la vista del sol/ y assi lo veen to
do el tienpo q̃ va subiendo y tot na decindiendo has
ta que viene ado su orizonte no descubre ento
do ni en parte la buelta del sol / y asi quanto mayor
fuere la parte que desta buel viere tanto ter
na mayor dia

¶ LICENÇ. Si debaxo del polo el sol se vee
seis meses continuos. Pregunto porq̃ ay alli
menos lumbre y menos calor   COSM̃

37   Ya sea declarado que nuestro o
rizonte descubre ciento yocheta
grados quees lamitad del cielo
y que quanto mas cerca del polo se abitare tan
to mas se vee labuelta que el sol da al mũdo
(quando anda de aq̃lla parte) demodo q̃ losq̃
abitan enlos circulos quando el sol entra en
el tropico labuelta q̃ el sol aquel dia haze
estoslaveran entera mente porque su orizõte
descubre toda la parte del cielo pordo el sol
este dia da subuelta: y asi por esta misma
Razon losque abitasen debaxo delos polos
loverian seis meses porq̃ su orizonte es lali
nea equinoçal Mas abnq̃ todo este tiempo
alli da lumbre el sol/por estar tan desviado
que quando mas se llega alpolo esta apartado
sesenta yseis grados y veinte y siete minu

tos de modo que avnq̃ alli seis meses continuos
se veria el sol por ser passados otros seis que nose
a visto y por la gran distancia que ay dealli al
sol menos escalienta ymenos ahumbra q̃ en
las otras partes Por que el frior continuo
Mas leuanta de vmedad quel sol puede cõsumir

38 ¶ LICENᴄ̃ Pues enbnas partes del mundo los
dias son grandes yen otras peq̃ños Pregunto si
entodo el año sebee el sol mas tienpo enbna par
te q̃ en otra ꜱ COS M̃

38

Vnque enbnas partes son los dias
mayores que enotras segun sea de
clarado Es de notar que elsol ensu
discurso emouimiento que entodo el año por
el mundo haze y qual cantidad de tienpo se bée
entodas partes Considerado lo que cadavna
parte tiene de dia eloque tiene noche Desta ma
nera Losque abitan debaxo dela equinodal
como tienen los dias yguales conlas noches cier
to es que el medio tienpo tienen dia yel medio
tienen noche Elosque tienen dia de quinze o

ras tienen dia de nueue oras porq̃ assi como
va creciendo el dia de doze hasta quinze assi
mesmo torna a menguar de doze hasta nueue
ylo mismo tienen de noche · y por Consigui
ente los que tien dia de veinte oras tienen
dia de quatro oras. Elo mesmo tiene denoche
Y los que tienen dia de vn mes sin noche tie
nen vna noche de vn mes sin dia E asi se puede
tener delos de mas De modo que mirado lo
que en cada parte ay dedia ecomo otro tan
to ay de noche se hallara ser tanto tienpo
del año el que elsol se vee enqual quier par
te como el que no se vee justa mente.

39 ¶LICENC Pues la lunbre dela luna es
causada por el sol y el sol tiene sienpre ente
ra lunbre Porque la lunbre dela luna crece
y mengua   ¶COSM

A Lgunos touieron que la luna
tenia lunbre desi mesma y que
quando estaua juntamente en
vn signo conel sol por elgran claror delsol

escurecia su lunbre y quanto mas se aparta
va del sol comencaua mas su lunbre apare
cer y quanto mas se allegaua menos parecia
esto no es assi porq̃ la luna no tiene nin
gun claror ni Resplandor suyo propio mas
el sol que le esta encima la alunbra eno cada dia
ygualmente hazia nos mas quando ella
esta derechamente hazia el sol alunbra la
parte de arriba y haze sonbra hazia la tria
y por esso nola vemos esto es en su conjuncio
y como por su mouimiento se aparta del sol
comiença vnpoco a resplandecer y parece
en modo de cuerno delgado y llamase enyric
to mono dies que quiere dezir de vndia y
como se aparta del sol mas rresplandor tiene
assi que alos ocho dias es llamada dia tho
mos que quiere dezir partida por medio
y tanto quanto mas deciende abaxo su rres
plandor tanto sube arriba su sonbra yalos
quinze dias es llamada Anphitricos que
quiere dezir llena porque es mas aparta

da del sol lo qual se prucua porq̃ quando elsol
senos esconde enel ocidente ella comiença apa
recer enel oriente. yentonçe toda su sonbra su
be arriba. yel Resplandor deciende hazianos.
y despues como se comiença de llegar alsol poz
aquel modo que se fue apartando. quanto
el Resplandor sube la sonbra deciende. yassi tor
na aboluer menguando enla manera que fue
creciendo.

¶ LICENÇ Pues la luna pasa el circulo
del zodiaco en veinte y siete dias yocho oras pz̃
q̃ dezimos q̃ cada luna tiene 29 dias ymedio. cõs.

Es assi que enveinte y siete dias y
ocho oras la luna passa el circulo
del zodiaco. mas enestos dias eo
ras no alcança al sol y por tanto passa adelan
te otros dos dias y quatro oras. hasta lo al
cancar y entonces se haze la conjuncion della y
del sol. y assi la luna dezimos que tiene vein
te y nueve dias y medio que se entiende deu
na conjuncion aotra. y los que hallaron el

conpoto dela luna Contaron las lunas aveinte
nueve dias e atreinta por que no acostumbraró
acontar menos devn dia entero y comiencan su
cuenta dende setiembre por ourra delos egipcios
q̃como es dicho entonces comiencan su año.

41 ¶LICENC De q̃ son vnas manchas q̃ ve
mos enla luna: COSM
41

Las manchas que vemos enla lu
na se causan enesta manera es de
saber que abnque el cuerpo lunar
natural mente sea oscuro/e desi propio como
sea dicho/no tenga lunbre alguna/es assi que e
algunas partes suyas Recibe mas lunbre yeno
tras es mas condensada· yenaquella parte q̃
es mas dispuesta tocada dela lunbre del sol
parece mas clara ylaque no es tanto parece
mas turbia o escura yasi parece aq̃llas machas q̃vemos·

42 ¶LICENC De q̃ se haze vn circulo q̃ ve
mos enderredor dela luna: COSM
42

Como la luna sea mas baxa elleya
da anos que ninguno delos otros

planetas y su figura es Redonda quando
el ayre no es muy escuro ni muy Resplandã
ciente toca la luna enel ayre consus Rayos
y haze enel figura Redonda yaquel circulo
no es lexos dela tierra mas nuestra vista
enesto se engaña que nos parece ser cerca de
la luna mas el ayre alla arriba estandel
gado que no se podria hazer porque las for
mas no se hazen sino en cuerpos gruesos
Y este circulo hazese mas amenudo quã
do haze viento de medio dia y quando
este circulo esta ygual mente debaxo dela
luna y se deshaze ensi mesmo muestra que
el ayre esta sosegado y quando va alguna
parte Roto muestra que de aquella viene
viento Los marineros lo tienen por señal
de tormenta oviento demasiado Este cir
culo tanbien se haze de dia pero no se vee
por la lunbre del sol:

+3 ¶ LICENC Dizese la luna es enla cabeça O
cola del dragon (que es drago) o como se entiende

+3

Val quier planeta· exçeto el sol
tiene tres circulos queson· Eqñ
te· Diferente· Epiciclo· El equã
te dela luna es circulo concentrico con la tie
rra y esta enla sobre haz dela ecliptica· El di
ferente es circulo excentrico y noesta enla
sobre haz dela ecliptica mas antes vna mi
tad suya declina hazia el septentrion y la
otra hazia el austro y corta el diferente
al equante en dos partes y la figura dela
cortadura se llama dragon· porque es ã
cha enel medio yangosta enel fin Pues
aquella cortadura por la qual se mueve la
luna de austro hazia septentrion se llama
cabeça de dragon y la otra cortadura por do
se mueve de septentrion hazia el austro se
llama cola del dragon·

44 ¶ LiCENÇ Deqse cavsa el eclipsi dela luna E
porqmas en vna opposiciõ qen otra?   COSM

r

Omo el sol sea mayor que la trra
es nessario que la mitad de la Re
oondez de la tierra sea sienpre a
lunbrada del sol/y la sonbra dela tierra es
tendida enel ayre haziendo se menor en
la Reoondez hasta que falte enla sobre haz
del circulo delos signos yn separable del na
dir del sol Nadir se dize vn punto derecha
mente oppuesto al sol enel firmamento.
de do se sigue que enel plenilunio quando
la luna estouiere enla cabeça o cola del dra
gon del nadir del sol oçerca y el sol estouie
re ala parte contraria dia metral mente
entonces la tierra se ynterpone enmedio
del sol y dela luna y la sonbra dela tierra
cae sobre el cuerpo lunar de do se sigue q̃
como la luna no tenga luz ni claridad sal
vo del sol desfallece dela lunbre e causa se e
clipsi quanto mas o menos dela sonbra

alcança y esto sienpre se haze en la luna llena
o cerca y como quiera que en qualquier ple
ni lunio no esta la luna puesta en la Cabeça
o cola del dragon debaxo del nadir del sol no
acontece en qual quier opposiçion eclipsi dela
luna

45 ¶ LICENÇ̃ Pregunto si los dos elementos
estremos pudieran estar sin otros en medio o
con solo un elemento en medio ⸱ ¶ COSM̃
45

Igo que no pudieran estar solos
los dos elementos estremos q̃ son
fuego y tierra ni p̃ tan poco conb
lo un elemento en medio lo qual se prueva
porque si no oviera otros elementos en me
dio quedara vazio entre el fuego y la tierra y
como no puede aver cosa vazia ⸱ el fuego dcin
diera a la tierra o la tierra subiera al fuego y
si la tierra subiera arriba como su natural sea
de abaxar O el fuego decindiera como su natu
ral sea subir yendo la tierra arriba y el fuego
abaxo toda la orden del mundo se deshiziera
tanbien si estos dos elementos de tierra y

fuego se tocasen sin aver otros en medio el fue
go entraria porlos aguxeros dela tierra y que
mar laya y tornarla ya en ceniza yassi ni
hobres ni animales niotras cossas avria y
si solo el ayre fuera en medio como tenga mas
natura de fuego que de tierra mudarase en fu
ego y si el agua sola fuera en medio por que
se llega mas a natura de tierra que de fuego tor
narase en tierra yasi la tierra llegara al fuego
y el fuego ala tierra como dicho es De modo
que bien convino segun las calidades destos
quatro elementos aver entre los dos estre
mos otros dos elementos medios

46 **¶LICENÇ** Que calidades tienen los eleme
tos E si pudiera aver otro medio entre ellos q
no fuera elemento.   **COSM**

46 **L**As calidades destos quatro eleme
tos son El fuego caliente y seco
El ayre Caliente y vmedo El a
gua fria e vmeda La tierra fria y seca Co
cuerdanse enel peso quela tierra pessa mas

que el agua El agua mas que el ayre El ay
re mas queel fuego · Y en quanto si pudiera
aver otro medio que no fuera elemento digo
que al poder de dios no se pone ningun ter
mino Mas si lo oviera criado · y no oviera
ayre con que el honbre alienta e menos de
aliento no puede biuir/ ni fuera agua de la
qual tiene honbre contino necessidad no o
viera honbres por lo qual fue muy conveni
ente y necessario que fuesen estos dos medios
entre los estremos para templança de todos quatro ·/

47 ¶ LICENC Pregunto si se mueven todos
estos elementos · E porque vemos vnos e o
tros no ·

COSM ·

47

D Estos quatro elementos los tres se
mueven es a saber El fuego el ayre
y el agua Mas la tierra que tiene por
lugar el medio de toda la spera assi como cen
tro ygual mente apartado del mouimiento de
todas las otras speras · E por su gravedad ·
e pesadumbre queda sin mouimiento y de

estos quatro elementos los dos que son gra
ves e pesados vemos es asaber la tierra y el agua
el fuego ni el are no vemos por q̃ como natu
ral mente son sutiles e ligeros carecen de cuer
po e nos vemos solo loque, tiene cuerpo tan
bien como nuestra vista es gruesa. yestos dos
elementos en su natural sinples son tan delga
dos que nolos puede la vista conocer /

48 ¶ LICENC Que cossa es el elemento del
48 fuego E q̃ especie de fuego tiene? ✠ COSM

El elemento del fuego es una muy
pura lunbre dela qual es aquel
circulo ospera que llamamos de
fuego que llega dela suprema parte delaire
hasta el cielo dela luna· y aunque el fuego
material que vemos tiene tres partes q̃ son
lunbre flama e brasa· ase de tener que este ele
mento es sola lunbre e no alguna delas otras
porque las otras an menester materia enq̃se
conserven· E delas species de fuego esta es la mas
singular /

L elemento del ayre segun sea dicho
es diaphano ensi·ꝓ ygual liibre q̃
del sol recibe. Enpero por acidente e
vna parte es mas claro /oescuro que en o
tras O mas caliente o mas frio /para lo qual
es desaber q̃ en este elemento o Region del ayre
ay tres partes q̃ son esta p̃mera mas cercana a
nos· clamedia /ela suprema Destas la suprema
parte comosea ynmediata alelemento del fue
go participa mas dela lumbre y calor de aquel
yaun del calor del sol· E esta primera parte q̃
es mas allegada anos por la Reberberacio q̃ los
Rayos del sol hazẽ enla tira escalientase De modo
q̃ la parte deẽmedio (q̃ llamamos media Regiõ del
ayre) como no participa de ningui calor yella sea ꝯsi
fria q̃da mas fria emas escura E por esto a esta
parte llamamos ayre caliginoso que quiere

dezir ayre escuro Enel qual por su frialdad
se engendran Las luuias nieues granizo e
piedra eotras cosas que vemos

C̶ L̶ ICENĒ Que cossa es Viento O co
mo se causa ⸝   ⸝ COSM̄

V iento es ayre que se mueue Rezia
mente Lo qual dizen algunos q̃
se causa delos vapores del agua y
dela tierra que el ayre Recibe ya contece q̃ algu
nas cossas menudas se ayuntan enel ayre
y lo vno enpuxa lo otro ya si haze el ayre
de rrezio mouimiento Tanbien dizen q̃ el
viento se haze del humo q̃ queda enla tie
rra enel estio despues que el sol espuesto
con mucho calor gran parte dela noche e
tita asi la vmedad dela trra y el agua y engruesa
se y hazese de mouimiento Rezio Lo mas cier
to es q̃ el viento se haze delos grandes moui
mientos que contino traen las mares de v
na parte aotra E assi vemos que enla mar

e cerca della ay mas Rezios e continuos
vientos que en otras partes dela tierra /

Si LICENÇ Si el mouimiento delas mares
causan los vientos pues las mares corre cada dia
Si porque cada dia no haze viento   COSM.

Todos los dias haze viento mas
no son sienpre tan grandes que
vengan anosotros ygual mente
Cada dia / tanbien muchas vezes avnque
do estamos nosentimos viento . no por eso
cessa que nolo haga en otras partes / tãbiẽ
Regiones ay que tienen vientos mas conti
nuos e propios que otras En pulla ay vn
viento que se llama arabolus. y en calabria
otro llamado japix y en francia otro llama
do tirois. y en vandalia o andaluzia otro
llamado solano E assi otras prouincias e
partes tienen vientos mas continuos e
Rezios que otras. lo quales o por estar cerca
de algunos estrechos de mar o por los sitios ẽ
qͣ las tales partes estan ÷

§2 ¶ Loicenc̃ Sie viento es ayre y el ayre
naturalmente es Caliente y vmedo por q̃ el vi
ento norte es frio y seco.   § COSM̃.

Ninguna cossa sebuelue mas psto
enotra que el ayre por que es
puesto entre vn elemento calic̃
te yotro frio y luego se buelue en natura de
aquellos y poresto quando esta cerca de
la tierra toma las Calidades dela tierra y q̃l
es la parte dela tierra donde el viento viene
tal es elviento y assi quando viene dela par
te ó Region de oriente la qual es caliente y v
meda elviento es caliente y vmedo ydela Re
gion de poniente que es fria y seca viene el
viento frio y seco y los dos cabos delatierra
es a saber norte y medio dia son frios eume
dos ya esta causa los vientos que dellos vie
nen son frios eumedos. Mas el norte a vnq̃
adonde nace es vmedo por Razon que quan

do viene anosotros haze viento claro llamamos
lo frio yseco. ychviento austro avnq assi mes
mo do nace es frio yvmedo enpero por que
viene a nosotros por medio dela torida zona
que es muy caliente calientase. ypor q viene
con ñublos cluvias aesta parte do estamos
llamamoslo caliente yvmedo.

53 **LICENC** Que Calidades tiene esta
nuestra Region — **COSM**

53

Esta Region enque estamos se
gun diversos vientos tiene di
versas calidades porque aqlla
parte que es cerrada por montañas hazia
oriente yocidente yes abierta hazia me
dio dia escaliente yseca yes buena para yn
vierno ycnverano es muy mala ypor lo cõ
trario yaqlla parte que es abierta aorie
te ydelas otras partes cerrada es caliente
yvmeda yes buena para en otoño ysitiene

lo contrario es fria y seca yes enotoño mâ
la yenla primâ Vera buena y esto q̃es di
cho delas partidas dela tierra se puede pro
var por las ventanas de cassa quelas q̃
son de parte de medio dia son malas enes
tio e buenas en ynvierno elas queson ha
zia el norte hazen el contrario y aesta causa
los antiguos hazian puertas enlas cassas
hazia el norte y medio dia enlas de medio dia
comian edormian en ynvierno yenlas del
norte el estio

¶ PILOTO

54 ¶ La Carta de marear tiene treinta y dos nôbres
de vientos. Pregunto si estos sehallâ entodas partes

¶ COSMOGRAPHO

54

E N qual quier parte que hôbre
este sehallaran estos treinta y
dos nôbres de vieñtos quela na
uegacion tiene porq̃ avnq̃ el elemento sea

vno damos nonbre al viento conforme ala
parte donde viene segun las diferencias de
los nonbres que por el mundo esu Redõ dez
son señalados enesta forma ∴ ÷ ∴

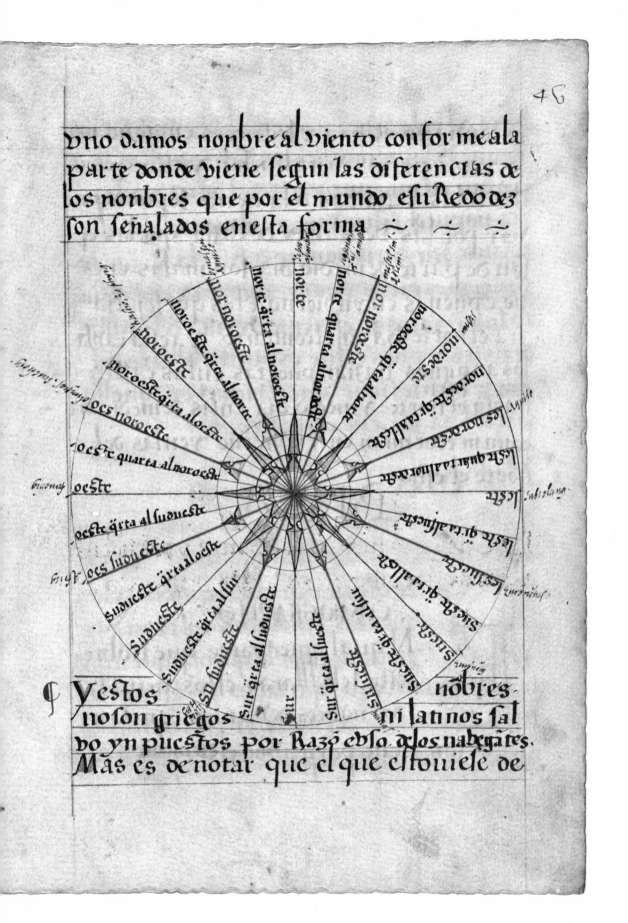

¶ Yestos nõbres
noson griegos ni latinos sal
vo yn puestos por Razõ ebsa delos nabegãtes.
Mas es denotar que el que estouiese de

baxo del polo artico no ternia leste nio este ni
alguno delos otros vientos sobre dichos Mas
por que en ninguna parte del agua y dela tie
rra no puede faltar viento· el viento que este
touiese sele daria nonbre de sur/ la Razon es
por q̃ ya es dicho q̃ es propiedad de este eleme͂
to subir ya si alli· el viento vernia dela parte
deabaxo y pues el sur espuesto debaxo del
norte bien se sigue que el viento· en aquella
parte sera sur eno otro alguno ÷

¶ PILOTO· Pregunto si se podra navegar cõ
algun viento hasta debaxo del norte E q̃ lõgitud
tiene el mu͂do en cada grado de latitud·   COSM

Igo que por qual quiera delos
vientos señalados enla carta de
marcar· e por la quenta que aq̃ di
re se podra navegar quitados ynconbenien
tes hasta debaxo del norte enesta manera/
todos los vientos ecepto leste E oeste llama

mos vientos de altura por que navegando
por ellos nos allegan o apartan del polo pu
es por qualquiera delos que son figurados de
de el leste hasta el norte sepodra navegar /
hasta el lugar que dicho es teniendo quen
ta q por cada viento q se navegare encada
cien leguas de camino sea parta delavia de
recha setenta y cinco leguas y por la media
partida treinta y siete leguas y media y por
la quarta diez yocho leguas ytres quartos/
pues mirada esta quenta dada vna buelta
al mundo epuesto nortesur conel lugar do
partio vera quanto esta apartado del norte
y aquello andara por la quarta medi partida
oviento segun el camino mas omenos fuere
siguiendose por la quenta que dicha es. y en
quanto para saber la longitud que el mudo
tiene encada grado delatitud es desaber que
quanto mas se va llegando al norte tanto
labuelta del mudo vañiedo menor porq como

el mundo sea cuerpo sperico quanto se va a
llegado alos polos va siendo su Redondez menor /
y aunque por qualquiera parte la Redondez
del mundo tiene. 3 6 0 grados entiendese q̃ los
grados quanto mas se van llegando alos po
los tiene menos longitud como parece por esta fig̃

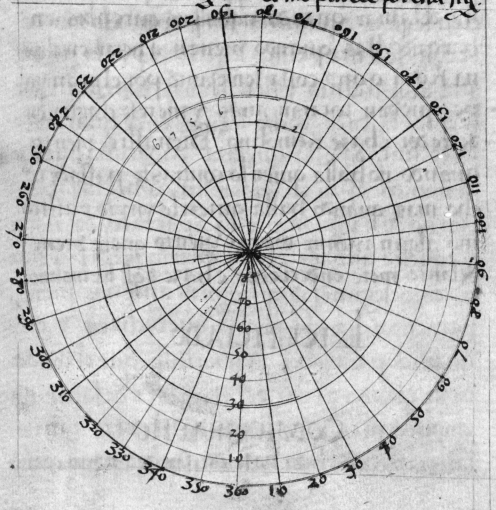

56 ¶ PILOTO. De que se haze el Remolino del

viento? COSM

Or semejança delas ondas dela
mar se podra conocer. porq̃ vemos
que las ondas van ala vera o ori

lla delamar quando nohallan quienlas en
bargue. Mas quando vienen a herir enalgu
na Roca o otra cossa semejante por el gran gol
pe quedan tornan atras y metese elagua en
derredor e haze Remolino. Bien assi el viento
quando no halla quien lo contraste passadere
cho mas quando halla quien le ympida assico
mo algun monte o otro Viento quele viene
delante mete enderredor e haze aq̃l Remolino.

¶ LICENCIADO:

57 ¶ Como se causan las luvias la nieve E granizo?

57 ¶ COSMOGRAPHO:

As luvias se causan devn ymor es

peso e vmedo del vapor dela tierra y del agua atra
y do por el sol y subiendo arriba condensase conel
frio e haze gotas menudas. E mientra mas sube las
gotas son mas gruesas. tan bien elayre conel frior.
dela tierra y del agua espessase y mudase ensustan
cia de agua y aesto llamamos ñublado. el qual qu
anto mas es peso tanto mas lubia causa La nie
ve se haze de aquel vmor vmedo q̃ sube dela tra
por el frior ques cerca dela tierra aprietase y espe
cesse se ecac hecho niebe. El granizo se causa
quando el vmor del agua yuesta assi como sea dicho su
be mas arriba y como alla arriba haze ayre
muy frio yela las gotas del agua etornalas
como piedra ycomo las gotas del agua son Re
dondas assi se haze el granizo Redondo. De mo
do quela primera condensicion es de agua La se
gunda de niebe. la tercera de granizo epiedra. E
assi vemos que acontece enfin dela primavera
caer granizo epiedra y eneste tienpo nocae ñi
eve mas sienpre cae en ynbierno. la causa es
por q̃ el vmor suso dicho enel verano sube mas

alto por el calor del sol y mientra sube enbuelve
se las gotas e hazense mas gordas eyelanse eha
zese el granizo ela piedra · Mas en ynvierno co
mo el frio sea cerca dela tierra asi el agua antes q̃
se engruese aprietase por la frialdad del tien
po e hazese la nieve ·

¶ LICENC̃ Porque enlas montañas altas
que son mas cerca del sol todo el año ay nieve
y no enlos valles ~   · COSM ·

58

Y A sea declarado que el ayre tiene
tres partes que son esta primera
y la media y la suprema · y el ayre
mientra mas arriba es mas puro y mas del
gado y mas frio enel medio · ycomo es mas
baxo e cercano ala tierra · es mas espeso y me
nos frio · porque como los Rayos del sol tocā
los valles elos lados delas montañas enopue
den yr mas adelante Reberberan alli ycalien
tan la tierra y por esta Razon el ayre que es
cerca dela tierra es menos frio ℰ por este fri
or del ayre tura mas la nieve enã ꝗllas partes
altas queno enlos valles ÷   ∵

59 ¶ LICENÇ Deq̃ se haze el arco que enel
ciclo parece quando llueve Edeq̃ se cavsan a
q̃llas colores.   § COSM·

59

EL yris o arco que entienpo
deluvias vemos se causa, dela
Reflexion delos Rayos solares
quando hieren enalguna nuve aguosa yca
usanse estando devna parte elsol ydeotra
la nuve· y como los Rayos delsol tocan la
nuve quiebranla yentonce parece aquel
arco· yalli donde Resplandece toma diver
sos colores Tanbien dizen que aquel ar·
co es ymagen delsol causada enla nuve·y
porque toda ymagen tiene la figura dequi
en Representa· yasi como elsol es Redõdo
parece aquel arco defigura Redonda y
assi como aquel arco noes sustancia con
viene saber cosa que este porsi mas y
magen desustancia bienassi noay colo

res mas semejança de colores. Otros dizen q̃
el arco es nube muy espessa y luziente q̃ tie
ne quatro principales colores delos quatro
elementos Del fuego color bermeja del ayre co
lor azul del agua color amarilla dela tierra
por las yervas e arboles color verde Tanbiẽ
dizen que dela mistura delas nubes y el aire
y el fuego y delos Rayos solares Resulta aq
lla variedad de colores que vemos ÷

60 ¶ LICENC ѕ̃i el arco se haze dela y mage̅
del sol (pues el sol es Redondo) porq̃ no vemos
el arco Redondo ÷   ѕ COS M̃

60

Omo el sol es mucho mas alto
que las nubes quando toca a
la nube de parte de arriba ha
ze en ella su y magen y por estar nosotros
en la parte baxa nola podemos entera mẽ
te ver mas tanto quanto el sol es mas cer
ca de oriente o de occidente tanto vemos ma
yor parte dela Redondez que el arco tiene ÷

E L trueno se causa del tocamien
to. delas partes del ayre q̃ corrē
vna contra otra Rezia mente.
y el Relanpago es vna parte del ayre que
se haze fuego er resplandece loqual se
haze enesta manera. El vmor vmedo assi
como es dicho sube arriba ycomo allega
ala suprema parte del ayre las partes de
este vmor corren la vna contra la otra
y del tocamiento hazen el Ruido del true
no y por este movimiento calienta se tã
to el ayre que passa en sustancia de fuego
yassi se haze la corruscacion orrelanpa
go que vemos quando truena. yavn q̃
todo se haze junta mente. mas ayna ve
mos el Relanpago que oymos el trueno
porque el Ver es mas presto que eloyz
tanbien se dize que la nuve aguosa co

ino sube arriba y se encuentra consu contrario
que es el fuego del topamiento se haze aquel
tronido y resulta el Relanpago ~

62 ℂ LICENᴄ̃ Deque se hazen las come
tas ~

62   COSM̃.

L A materia deque se engendran to
das las cossas enlos elementos es
endos maneras. es asaber Materia
Caliente y seca quese llama exalacion yotra
caliente yumeda que se llama vapor Desta
Caliente y seca se engendran las cometas e
esta manera. quando la exalacion sube hasta
la spera del fuego quema se alli delo qual ve
mes aquel fuego. E porque ay planetas q̃
atraen assi algunas exalaciones dizese ser
dellas el que se vee enel cielo. como camino
de nubes que vulgar mente se llama cami
no de sanchiago ~

63 ℂ LICENᴄ̃ De q̃ se causan los Rayos y
que efectos hazen ~   COSM̃.
63

L Os Rayos se causan. oson en
dos maneras. La vna es parte

125

de ayre que se haze fuego y biene subita
mente La qual se haze assi quando se ha
ze aquel Ruydo del trueno eRelanpago q
dicho es por aquel mouimiento de aqllas
partes de vmor vna contra otra la prina
pal parte sube y recoge todo el fuego q
esta derramado eluego deciende ecorre a
vna parte eaotra con muy gran fuerça has
ta que halla cossa tan grande qla contraste
y estos Rayos queman las cossas altas
y hazen grandes efectos como adelante
dire Otro ay que es de sustancia de pie
dra el qual se causa enesta manera Quãdo
el vmor vmedo sube arriba esube con al
guna cossa que tiene sustancia de tierra
por el calor del sol condensase ebueluese en
piedra y tornase enla concavidad dela nu
ve hasta que la nuve se parte y entonce
la piedra deciende y hiere enlo que halla
estos abren las torres ye dificios ehien
den naos ehoradan la tierra ehazen o

tras cossas grandes ·Destos Rayos grādes
efectos vemos Enespecial que al honbre
muerto con Rayo secā los huesos y no pene
tra el cuero Enel madero dolado hunde el
clavo yel. madero queda sano. Regala la can
pana y no quema la soga· Ronpe e quema la
cuba enose vierte elvino por tres dias ·÷

¶ PILOTO

64 ¶ Que s la mar E porque se llama Occano ~

¶ COSMOGRAPHO

64

L Amar es ayuntamiento delas
aguas ycs el mismo elemento del
agua llamase occano de okis é
griego que quiere dezir presuroso los grie
yos elatinos por este nonbre occano lenō
bran· por la presteza ca presuramiento cō
que corre Estemar todos los dias noes y
igual ensus moui mientos· antes por siete
dias sealça creciendo ypor otros siete seRe
trae menguando esu color pare mudarse
segun la variedad delos vientos·÷

A sabiduria de dios vio que sin
calor eumor ninguna cossa podia
biuir ycomo la tierra es fria yse
ca y puesta sobrella la fuente del sol eporque
consola calor nopodia biuir cossa alguna pu
so la tierra en medio dela fuente del umor q̃
es lamar para tenplança del calor del sol
yassi por estas dos cossas se tienpla deuna
parte edeotra Desta mar loque viene hazia
ocidente haze dos partes la una hazia me
dio dia y la otra hazia el norte por los dos
lados dela tierra y semejante mente hazia
oriente otras dos partes y quando los dos
braços dela mar de ocidente se encuentran
del gran golpe tornan las mares atras
elas unas van eotras vienen bien assi los
otros dos braços se encuentran enla otra
parte ehazen lomismo yassi causan el
crecery menguar Otros dizen quelas al

turas e grandes honduras que son dentro
dela mar hazen boluer las mares de vna par
te a otra Lomas cierto es q̃ la mar que es el
elemento del agua haze su mouimiento co
mo los dos elementos de fuego y ayre natu
ral mente y este mouimiento viene tan cõ
for me conel crecer y menguar dela luna q̃
algunos dizen que la luna cause el crecer y
menguar dela mar / por que las mareas ocre
cientes justa mente con la luna vienen q̃
si como la luna nace y se pone cada dia en diuer
sas oras assi a las mismas oras vienen las
mareas

66 PILOTO Por que la mar es salada

COSM·

66
LA mar es en mayor cantidad pu
esta debaxo dela torida zona /
que es aquella parte enq̃ el sol es
mas que mante E por el gran calor del sol q̃
le estan cerca loque del agua es puro eligc
ro arrebata cconsume el calor del sol y los

enbates delos vientos e queda topessado e
terrestre. Ela mar espesase. ehazese amar
ga osalada. quevisto es que el agua setor
na salada quando gran calor desol leda en
cima. Eassi mas salada sedize hallar lamar
enel otoño muestro que en otro tienpo por
Razon del calor del estio quesobre ella æ
passado

67 ¶ PILOTO Lamar sies Redonda Ollana
67 o que figura tiene? ʃ COSM

Elagua delamar es Redonda y
que sea assi parece por dos Ra
zones. La primera como el a
gua sea cuerpo devna forma onatura y
el todo consus partes es devna forma mis
ma. Assi cada vna delas partes tiene la for
ma del todo. Pues las partes delagua na
tural mente desean la forma del todo assi
como vemos enlas gotas del agua ǭ caen co
mo Rocio enlas yerva aqueson Redondas
assi quelas partes muestran la figura del

todo. La segunda poniendo vna señal e
la Ribera dela mar y salga vna nao del pu
erto y tanto se aluengue dela tierra que la
vista del honbre que estouiere cabe el pie
del mastel estando queda la nao no podra ver
la señal Mas la vista del mismo honbre si
se sube encima del mastel vera la señal avnq
estando abaxo cabe el pie esta mas cerca de
la señal que encima yesto causa por ser Re
donda la mar ~

68 ¶ PILOTO Si vn piloto por la mar per
diese la carta de marear y el aguja Como se Re
gira para hazer su viaje ~   COSM.

68

Vnque en tal Casso son necessa
rios algunos particulares abi
sos aqui breue mente digo ꝗ el
piloto deve presu poner dos cossas La vna
la derrota que llevava segun el yntento de
su camino La 2ª en que tiempo le aconteçe
la tal falta por ꝗ conforme a esto el sol le
sirba por aguja e derrota ⸆ conoscimiento

de bientos lo qual entienda enesta manera /
A los onze de março el sol sale al leste y se
pone al oeste yes desaber q̃ dende onze de mar
ço hasta onze dejunio queson tres messes q̃
el sol tarda enllegar dende la equinocial hasta
el tropico de cancer se aparta dela dicha line
a· veinte ytres grados ytreynta ytres minu
tos· los quales sube enesta manera el pri
mer mes onze grados y quarenta y siete mi
nutos yel 2º· siete grados y ãnq̃ ydos minu
tos y el 3·· tres grados y ãnq̃ y ãnco minutos
e Repartidos 3 6 0 grados dela Redondez del
mundo enlos treinta ydos bientos dela nã
vegacion cabe devn biento aotro onze gra
dos y quinze minutos demodo que apar
tando se elsol veinte ytres grados ymedio del
leste hazia la parte del norte apartase dos
Runbos yassi diremos que estando el sol del
tropico de cancer sale al les noã oeste epone
seal oes noroeste y por el modo que el sol
hizo su acesso dela linea al tropico enestos

132

en estos tres meses torna a bazer su Recce
so en otros tres del tropico a la linea. E a los
treze de setienbre torna a salir al leste y po
nerse al oeste. y por esta manera se cotara
los otros seis messes del año q̃ el sol anda a
la parte del sur los tres que abaxa dende la
linea al tropico de capricornio E los tres q̃
tarda en bolver dende el tropico a la linea :
Pues conforme al tienpo en q̃ le acaece la
tal falta sabra el Runbo en q̃ el sol nace e
se pone. y es de notar q̃ como las oras del dia
natural son . 24 . de los vientos ocho en cada
tres oras el sol anda un viento y en ora y
media medio viento. y en tres quartos de o
ra una quarta pues mirada esta quenta por
el Relox cierto conocera . en cada ora del dia
en q̃ Runbo le esta el sol en esta manera. en
onze de março el sol sale a las seis al leste
y a las tres oras del dia estara al sueste y
al medio dia estara al sur y a las tres de la

tarde al sudueste y alas seis sepone al oeste · y
conforme àesta Regla poora Repartir las me
oias partioas y quartas · ya ssi sabra el camino
o oerrota quea de llevar · Yten sea denotar que
quantas mas oras tiene eloia · tantos mas vie
tos anoa el sol deoia y menos denoche · y por el
contrario · quanto es mayor la noche tanto
menos vientos anoa el sol de oia y mas deno
che Demodo quesi el oia tiene ooze oras qua
tro vientos anoa el sol de oia e quatro de noche
esi el oia tiene quinze oras · cinco vientos an
oa de oia y tres denoche · ysiel oia toviere
nueoe oras/tres vientos anoa de oia y cinco
de noche · E assi poora contar los Rumbos q
el sol anoa oe oia · eoe noche segun fuere el
tienpo ylugar donde estoviere ·

¶LICENÇ°
69 ¶ Como esta la tierra sitnaoa enel mundo ·~

¶COSMOGRAPHO ·
69
E L elemento oela tierra es puesto
en medio del mundo y por esto mas

baxo porq̃ como es declarado el mundo es Re
dondo y toda cossa Redonda el medio es mas ba
xo y q̃ la tierra este en medio del firma mē
to e octaua spera assi semuestra porq̃ a los
que estamos enla superficie o sobre haz dela
tierra nos parecen las estrellas de vna mis
ma cantidad en qualquier parte del cielo
q̃ esten ora enel medio / o en oriente o en occidete
y la causa desto es porq̃ la tierra ygual men
te dista o se aparta delas estrellas siguese
luego que esta en medio del firmamento y
ten q̃ i la tierra por alguna de sus partes estoui
ese mas llegada al firmamento que por otra
el q̃ estouiese en aq̃lla parte que mas estouie
se llega al firmamento no veria el medio cielo es
to es contra ptolomeo y todos los philosopos q̃
dizen que do quiera que honbre este sienpre
nacen conel seis signos y sele ponen otros seis
y el medio cielo vee y el otro medio sele asconde
luego ygual mente dista del firmamento y
es Redonda y es centro o punto dese mismo
fir mamento ꞏ∻

¶ LICENC̃ Pues los elementos se cercan v
nos aotros pregunto sila tierra es debaxo o
encima del agua :

COS M̊.

Val quiera delos tres elemẽtos
cerca enderredor latierra salvo :
quanto lasequedad dela tierra Re
siste la humidad del agua para vida delos
onbres ç animales q̃ biuen por Respiracion
Algunos dizen qued agua nocubre de toda
parte la tierra por Razon dela ynfluencia de
ciertas estrellas queson cerca del polo artico
lo qual parece por que el polo artico atrae
assi las cossas secas yel polo antartico las
cossas vmedas pruebase enla piedra ymã
que por vna parte enseña el polo artico por
la qual atrae el hierro ç por la otra parte
huye ÷

71 ¶ LICENC̃ Pues la tierra es en medio dela
yre yespessada por q̃ no cae pues toda cossa de
pesso puesta enel ayre cae ÷

COS M̊

Anque el ayre cerca la tierra de
todas partes la tierra no tiene don
de caer cano ay otra cossa mas ba
xa donde se sostenga por que como ella sea
mas baxa no puede abaxar mas y ninguna
cossa la sostiene avnque algunos dixeron qla
tierra se sostenia enel ayre como la nao enel
agua Otros dixeron que como el fuego cerca
la tierra enderredor detodas partes conla
fuerça que tiene de tirar arriba la tierra
nose puede mouer ni arriba ni abaxo ni a
ninguna parte a semejança delo que se dize
dela sepoltura de mahoma ques dehierro cer
cada de piedra yman yla virtud natural de
las piedras sostienenla enel ayre eque assi la
tierra es en medio del elemento del fuego y
consu virtud el fuego detodas partes la sos
tiene La verdadera Razon es que como to
da cossa graue e pessada va alcentro por q
el centro es punto del firmamento ecomo la
tierra sea degran pesso egraveza naturalme

te seva al punto del firmamento. y todo lo
q se mueve dende el punto sube. pues si la
tierra se mouiese. de en medio hazia la cir
cunferencia subiria. loqual es ynposible.
y assi la tierra es en medio del mundo y no
se mueve ni puede mouer /.

7² ¶LICENC Pregunto si la tierra es Re
7² donda ollana o que figura tiene.   COS.M.

Aticrra es Redonda avnq algu
nos creyendo mas ala vista qala
Razon dixeron que era llana (mas
no es assi porque si la tierra fuese llana las
aguas delas luvias que caen enla tierra elos
Rios nocorrerian mas ayuntarse yan en vn
lugar hechos laguna obalsa. yten/estrellas
ay que parecen en vn clima eno parecen en
otro. los de egipto veen vnestrella llamada
canopus que nosotros nola vemos y esto
no acontecería si la tierra fuesellana. yten
quela tierra sea Redonda detodas partes pa

sism

rece porque los signos y estrellas no nos
nacen ni se ponen ygual mente a todos los
honbres en qual quier parte que esten. mas
primero nacen y se ponen a vnos que a otros
y esto causa la hinchazon o Redondez dela tira
y assi vemos que vn eclipsi dela luna no y
gual mente en vn tienpo parece a todos. y
que la tierra sea Redonda dende el septentri
on a lauftro y por el contrario mueftra se asi
por q los que eftamos hazia el septentrio
nos parecen sienpre algunas estrellas que
estan cabe el polo artico y otras estrellas q
estan cabe el polo Antartico nunca las ve
mos pues luego si alguno fuese dende el
septentrion hazia el austro tanto podria
andar que las estrellas que primero vcya
sele escondiesen eno las viese y quanto
mas se llegase al austro tanto menos veria
las estrellas questan cerca del septentrion.
y entonce veria las estrellas del austro que
primero no veya que estan cerca del polo a
tar tico. Y a la contra acacceria el que fuese

dende el austro al septen trion. Ela causa es so
la mente la Redondez dela tierra ÷

73 ¶ LICENC. Como es la tierra Redonda ÷
pues ay enella tantos valles hondos y montes
que suben sobre las nubes.   COSM·

73

Vestra pequeñez lohaze quelas
cossas chicas nos parecen grandes
sino dezime quan grande puede
ser aquel monte olinpo contoda la tierra aco
paracion del cielo· cierto es muy poco o no na
da· pues pongamos casso que aqui ay tierra
que oviese des depassar en vn passo y esto ni
ese poca cossa mas alto vn pie que otro cierto
es que no os pareceria que avia alli valle ni
montaña· Mas silo oviese de passar vna cossa
muy pequeña parecer le ya vn gran valle pues
bien assi los valles ni los montes en respeto
del cielo no quitan la Redondez dela tierra ÷
Mas es desaber que alguna cossa se dize Redo
da en dos maneras vna Regular y otra y Re

gulares es quando/las lineas traydas del centro
a la circunferencia q̃ son yguales y en esta ma
nera no es la Redondez dela tierra Otra ma
nera se dize Redonda y Regular y es quando
todas las partes noygual mente distan del
centro y en esta manera es la Redodez dela
tierra :

74 Ⓛ ICENℂ̃ Que tamaña es la tierra en Res
pecto del cielo ·   COSℳ̃

74

Ⓛ A tierra es tanpoca en Respecto
del cielo que es cassi ynsensible
lo qual se muestra por esta Razõ
La menor delas estrellas del firmamento ala
vista notables · es mayor que toda la tierra y n
es vna estrella en Respecto del cielo es cassi
nada · por que en solo el medio cielo q̃ vemos
ay tantas estrellas y tanto lugar del cielo sin
ellas · y pues es assi quanto sera menor la tira
que el cielo pues es menor que vn estrella y q̃
la tierra sea tanpoca en Respecto del cielo pru
ebase por el sol que estando denoche debaxo

de nuestro hemisperio da e comunica su liubre atodas las estrellas q̃ estan sobre nos sin que la Redondez de tierra y agua que es tan en medio ocupen ni ynpidan a su luibre cossa alguna Enlo qual se demuestra quán poca cossa es la tierra en Respecto del cielo

❡ LICENC̃ Si la tierra fuese horadada de vna parte ala otra por el centro y se echase vna piedra por aquel aguiero · Preguinto donde pararia

COSM̃

Declarado sea que la tierra es mas baxa que ninguno delos otros elementos y por esto el medio de ella q̃ es el centro · es mas baxo que ninguna delas otras partes ꞏ pues aquella piedra que honbre echase por aquel aguiero decendiria hasta la mitad dela tierra a do es el centro y naturalmente alli se sosternia por que si mas andouiese subiria y lo pessado naturalmente no puede subir

142

76 ¶LICENÇ. Como se causan las fuentes en
la tierra E porq̃ vnas aguas son dulces eo
tras saladas ~   S. COSM.

76

Ya sea declarado quela mar es ele
mento del agua y della salen
todas las aguas yaella bueluen
en escripto es que las aguas tornan al lugar
conde salieron por que puedan otra vez correr
Pues como las aguas corren por las conca
vidades dela tierra, si al cabo dela concauidad
hallan dureza conque no pueden passar ade
lante ni puede tornar atras por la fuerça de
la otra agua que viene assi hierue sobre la ti
erra e haze fuente grande ochica segun la cãti
dad del agua que corre Ccomo el agua es de
leznosa e passa ese cuela por los agujeros de
la tierra dela tierra por do passa toma diver
sos sabores porque si passa por lugar are
noso opedregoso toma sabor dulce E si pa
ssa por tierra salada toma el sabor salado E si

passa por tierra lodosa sale agua de mal sabor
E si passa por piedras de cufre o de cal o de alum
bre es amarga Cassi segun la diversidad delas
tierras toma el agua diversos sabores

77

Vnq̃ el agua delos poços viene aellos
en aquella manera que es dicho delas
fuentes el poço no se hinche ni el a
gua Rebosa porque no tiene termino donde
pare ca el agua halla por do correr cassi pas
sa se que no bierue arriba Eque alos poços
venga el agua por las cavernas dela tierra pru
ebase porque cerca delos Rios sienpre se ha
lla agua e si se haze vn poço cerca de otro el
agua corre e passa del vno al otro Eavnq̃ ay
algunos poços que no tienen agujeros ni ca
vernas por do les venga agua por estar en lu
gares altos e secos y no ay falta de agua ene

144

llos· esta se causa dela misma tierra que a
vnque natural mente sea seca tiene vn v
mor por acidente que aguisa de sudor cae
gota agota· enel pozo· loqual se prueua q̃
sea del sudor dela tierra por q̃ enlos lugares
secos sethallan pozos conagua·

78 ¶ LICENÇ Pues la tierra y el agua natu
ral mente son frias porque en ynbierno el
agua delos pozos sale caliete y beranofria·

COS M

78

De suso esdicho que conel frior del
ynbierno se cierran los poros de
la tierra ycomo el humo della
nopuede salir eque da dentro escalienta el
agua mas enberano la tierra conel calor a
brese yel humo sale yamenguada la calor el
agua sale fria· Mas la Razon principal es
porque toda cossa sealegra desu semejante
y huye desu contrario· En ynbierno el ayre
frio toca la tierra· y la calor açidental dela

tierra huye hazia dentro yesconde se y por es
to el agua del pozo en ynbierno se calienta mas
enberano la tierra de encima se gina por el calor
del sol y el frio dela tierra huye hazia dentro y
haze el agua fria y enberano el agua dela fuente
es menos fria qla del pozo porques mas cer
ca deluyre yla del Rio mas caliente por la mis
ma Razon:

79 ¶ LICENÇ Dizese que ay gentes q se lla
man antipedes q andan en la otra parte dela
tira debaxo denos Preguto como andan de
baxo denos   ⸱ς COS M.

79 COmo la tierra sea Redonda segun sea
prouado los q abitan dela otra parte
dela tierra en nro opposito estan sus
pies contra los nuestros a semejança de quando
honbre esta cerca de algun lago de agua que en
el agua vee su sonbra con la cabeça ala otra parte
elos pies contra sus pies: pues asi aquellas gen
tes llamamos antipedes que quiere dezir que
tienen sus pies contra los nuestros y assi tie

nen ellos sus cabeças derechas al cielo co
mo nos. De modo que no sea de ymaginar q̃
nosotros estamos encima dellos ni ellos en
cima de nosotros porq̃ el cuerpo Redondo
no tiene parte de encima propia mente. y
assi entre sus pies y los nuestros es la Re
dondez de tierra y agua. Y es de saber q̃ noso
tros y ellos no tenemos en vn tienpo Vera ni
ynbierno ni los otros tienpos del año. mas
quando nosotros tenemos verano ellos tie
nen ynbierno y por el contrario y quando no
sotros tenemos dia ellos tienen noche por
que en vn tienpo no alunbra el sol a nos y a e
llos y a esta causa. tenemos diferentes los tie
pos porque como el Resplandor del sol cause
el dia y la sonbra dela tierra la noche quando
el sol es arriba en nuestro hemisperio la so
bra es en el hemisperio de abaxo y por el co
trario Cassi ninguna ora ay q̃ no aya noche e
dia por la Redondez dela tierra

¶ LICENC̄ Que cossa es hemisperio?

Hemisperio es la vna mitad del
cielo: E assi los hemisperios son
dos es asaber hemisperio alto e
hemisperio baxo El hemisperio alto es esta
mitad del cielo que nra vista descubre por o
rilla o entras dela tierra e llamamos le alto
por q̄ en nro Respecto es encima dela tierra q̄
vemos. y el hemisperio deabaxo es la mitad
del cielo que no vemos y hemisperio sedize de
semi spera· es asaber la mitad del spera —

¶ LICENC̄ Que cossa es clima· E quantos
son los climas e que nonbre Eanchura tiene ca
da vno ÷

Clima es vna linea trayda de orie
te a ocidente ygual mente apar
tada dela equinocial· y es de sa
ber quelos antiguos dimidiero aquella tie
rra deque touieron noticia ser abitable ensie

te partes oclimas las quales sola mente se
ñalaron dela eqninocial ala parte del polo
artico E dieronles nonbres de cibdad monte
o Rio O otro lugar famoso por do se señalan·
Yestos climas noson yguales· en longitud
ni la titud· porq̃ enla longitud la Redondez
dela tierra· seba angostando dela linea eq̃
nocial alos polos (segun dicho es) ya si la lon
gitud del primero clima es mayor q̃la del se
gundo· y assi delos otros· Los nonbres eanchu
ra decadauno destos climas eseste

El prim clima sellama diameroes del nõbre de
meroe ysla enel Rio nillo Comiença de nde doze
grados y quarenta y cinco minutos delaeq̃no
cial y llega hasta los 20 grados ymedio es el
anchura deste clima sietegrados y quarenta y
cinco minutos

El 2º· clima sellama dia siene del nõbre de sie
ne cibdad de egipto Comiença delos veinte
grados y medio y llega hasta los 27· grados

y medio es el anchura deste clima siete
grados ~

El 3 clima se llama dia alexandros del non
bre de alexandria cibdad de africa Comiē
ça delos. 7.7. grados y medio y llega hasta
los 33. grados. y. 40. minutos. es el anchu
ra deste clima Seis grados y diez minutos :

El 4. clima se llama dia Rodos del nōbre
de Rodas ysla Comiença delos 33 grados
y 40. minutos y llega hasta los 39. grados
es el anchura deste clima cinco grados y
veinte minutos. ~

El 5 clima se llama dia Romes del nōbre
de Roma vrbe Comiença delos 39 grados
y llega hasta los quarenta y tres grados y
medio es el anchura deste clima quatro gra
dos y medio.

El 6 clima se llama dia borestenes del non
bre de vn Rio de scithia Comiença delos
quarenta y tres grados y medio y llega has

ta los 47 grados y quinze minutos es
el anchura deste clima tres grados y qua
renta y cinco minutos ~

El 7º clima se llama dia Rifcos del nôbre
de vnos montes de alamania Comiença de
los 47 grados y quinze minutos y lleua
hasta los çinquenta grados y medio es el
anchura deste clima tres grados y quinze
minutos ~

Es denotar que Cada vn grado destos tie
ne diez y siete leguas y media de camino e
por esta quenta se sabra que anchura tiene
cada clima E contado 60 minutos por grado.

17 legas

E avnque allende destos climas ay tierras
yslas E abitaciones de gentes Estas porq
tobieron ser de abitacion yntenperada no las
contaron debaxo de clima ~ Mas por expe

riencia vemos que debaxo dela equinocial
que tovieron por yn abitable creyendo aver
alli gran calor agora tierra abitable e muy

tenplada hallamos y la Razones porq̃ como
yasea dicho ser alli los dias enoches yguales lo
que el sol de dia escahenta tienpla el frior de
la noche de modo que todo el año ay cassi v
na continua tenplança ÷Mas encadavno de
los climas diferentes noches edias tienen segũ
de suso sea notado ÷

¶ PILOTO

¶ Enla navegacion delamar sepraticã·derrota· e
Runbo· Pregunto q̃ cossa son ÷

COSMOGRAPHO

E S denotar que enla navegacion
delamar separatican especial m̃ẽẽ
estos tres nonbres es asaber Viento
Runbo· e Derrota· El viento yasea declarado q̃ ti
ene treinta ydos diferencias denõbres los q̃les
se dividen entres partes queson Ochovientos
principales yocho medios vientos y diez yseis
quartas Los ochoviẽtos p̃ncipales son estos·

Estos vientos en lengua de ytalia se llamã asi.

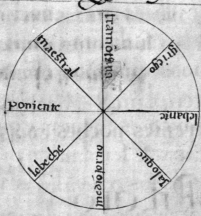

Los otros ocho que se llaman medios vientos.
O medias partidas son estos ynfra escriptos. los
quales se llaman assi por que son medios entre los
ocho principales cuyos nonbres so estos. ÷

Entre estos. 16. vientos ay otros 16. q se llamã qr tas.
los quales tiene nõbre delos vientos colaterales. segu
enla pregunta. S 4. sea declarado ÷

Runbo es aquellas Rayas oviruulas conq̃
se señalan estos vientos enlas cartas demarear.
los quales Runbos son como carateres (decada y
no delos vientos) E demostracion dela parte don
deVienen eadondeuan. E porq̃ seui sea declara
do ay vientos emedios vientos equartas. por
tanto los Runbos se señalan enprietos. Verdes
e Colorados. los p̃etos denotã los. 8. vietosp̃ncipa
les. los verdes los medios. Los colorados las quartas.
Demodo q̃ porestos Runbos. los vientos sedistin
guen econoscen ~

Derrota es el Camino q̃ selleua porlamar na
vegando lo qual (seguin el derecho tiene vn vieto
con otro) alli nonbramos la derrota enesta mane
ra. El q̃ estouiese entoledo equisiese yr aseuilla y
ria por derrota de norte su. q̃ seentiende llevando
el norte enlas espaldas. yla cara al sur omedio
dia Mas si fuese de toledo agranada su derrota se
ria noroeste sueste. que es llevando la cara alsu
este. y tras el lado ysquierdo el norte. Demodo q̃
seui elsitio ẽq̃ las partes esta colos vientos. tales
la derrota que se lleva enla navegacion ~

Finis.

Part Three

THE BOOK OF COSMOGRAPHY
Translation

BOOK OF COSMOGRAPHY
in which is given a description of the world
addressed to H.M. the Emperor
Don Carlos our lord,
made by Pedro de Medina,
Cosmographer

SACRED IMPERIAL CATHOLIC MAJESTY:

Experience, the mother of all things, tells us the extent of profit and utility of knowledge to be found in writing. The science of cosmography suffices to give proof of that; here are three major reasons. The first is that through it we achieve a better knowledge of the great power and wisdom of God, because it treats of the heavens and other great marvels that are in the world. The heavens, says the Royal Prophet [David], extol the glory of God, and Saint Gregory [says that] the marvelous works which we see created as a testament and sign are from our Creator. The second is that the study of cosmography enables us better to understand the divine scriptures because in them the universe and inhabited parts of the earth are treated again and again. The third reason to study cosmography is that by it we are introduced to and have explained to us the books of natural philosophy as well as those on the creation and decay of heaven and earth and meteors, and even an understanding of the books of the poets.

Now, as I wished to aid those who want to know something of that science, and in particular to advise sailors upon the seas, I gathered my energies to compose a brief description of the world, of which I have written and compiled in this book. I have organized it into questions put to a cosmographer by a graduate [*licenciado*] and a pilot, which seemed to me the most suitable style for easy comprehension of what is said. Moreover, it seems proper that, as this book treats of such majesty as the monarchy of the universe, it should be offered as I herewith offer it [to your Majesty]; and I humbly pray that it be accepted and commended, or—as our Lord with his eyes beheld the small gift of wisdom from his servant and found it good—not measured by the smallness of my work but by the great good will with which I hope to serve.

Table of Questions Which Appear in This Book

BOOK OF COSMOGRAPHY in which is presented a very useful description of the world, that is, of the heavens and stars, the sun and moon and the elements. It is arranged by questions put by a graduate and a pilot to a cosmographer. It begins in this way:

1. GRADUATE: As we are to treat of cosmography, I ask what is cosmography and how it came to be so called.

COSMOGRAPHER: Cosmography is the description of the world: that is, of the *cosmos*, the Greek for world, and *grapho*, for description.

Cosmography is thus a description of the world. And in this description are included geography and hydrography. Geography is the description of the earth, so called from *geos* which is earth. Hydrography is the description of the sea, from *hidros*[1] which is water, so that we are engaged in describing heaven and the elements of which the world is made.

2. GRADUATE: What is the world, why is it so called, and of what parts does it consist?

COSMOGRAPHER: The world is the universe of men made up of heaven, earth, the sea, and other elements. It is called world [*mundo*] because it is always in motion [*movimiento*] and no rest is ever granted it. This world is divided into two parts or regions, to wit: the celestial region and the region of the elements.[2] In those two regions are fourteen simple bodies which are: First the earth, which is the center of the world—dry, black, heavy, and dense. The second is water—cold, heavy, and translucent; the third is air— wet, volatile, and light; the fourth is fire—hot, light, volatile, and luminous. Beyond these four elements, which are the primary ones of all existing things in nature, there are ten other bodies called heavenly bodies, each of which circumscribes another until the tenth envelops all and is itself not circled by any other body. The first of these heavens is the circle of the moon; and beyond it is Mercury; the third is Venus, and the fourth the sun, the fifth Mars, the sixth Jupiter, the seventh Saturn; the eighth is the place where are all of the fixed stars which make up the firmament, and below that all circles are transparent, because we can see the stars which we would otherwise not see. Above the eighth is the ninth or crystalline sphere, luminous as crystal, in which there is not a single star, and it is called primum mobile, the tenth is the empyrean which envelops all. It is immobile and consists of pure light.

[*See figure, page 47*]

3. GRADUATE: Do the nine spheres move and do they move together or separately?

COSMOGRAPHER: The nine spheres[3] move in two different ways. One is that of the primum mobile from east to west, returning again to the east. The other eight lower spheres move in a contrary direction from west to east. This first movement with its impetus and swiftness provokes all the other movements and makes them turn once around the earth in a day and a night; and that is called a "snatching" or forced movement. Some say that it should not be imagined that the primum mobile moves the other lesser spheres by abrupt or strong force of attraction because so abrupt or forceful a movement could not emanate from heaven, since in heaven there is no resistance, and also a round body cannot attract another outside itself. If those two have no rough surfaces, and the heavenly bodies are round and have no rough surfaces, they say that this movement of the lesser spheres is due to some property which the primum mobile activates in them or to some other natural cause, such as that for instance which makes naturally heavy bodies rise up to form a vacuum. Such a movement, however, is not due to a force nor is it natural, but rather it belongs to the miraculous.

4. GRADUATE: How it is proved that the heaven moves from east to west.

COSMOGRAPHER: Because we see that the stars gradually rise in the east one by one, until they reach the center of the sky, maintaining the same distance and orientation among themselves; and we then see them descend uniformly until they set in the west. So, for instance, we see that those stars close to the Arctic Pole [north celestial pole], which are never hidden from us but always visibly moving, run their circles near the pole from east to west in the same manner and always keep the same distance from one another. From this description of the stars which rise and set as well as of those which we see all the time, it is proved that the firmament or eighth sphere moves from east to west.

5. GRADUATE: I ask whether the sky is round or what shape it has.

COSMOGRAPHER: That the sky is a sphere is proved by three arguments. The first is analogy,[4] the second utility, and the third necessity. The first is analogy because the sensible world is made to resemble the archetypal world which has neither beginning nor end, and because of this resemblance the sensible world has a round or spherical shape which does not permit the determination of beginning or end. The second reason is utility because, of all isoperimetrical forms, the sphere is the largest body and of all forms the round form has the greatest capacity; and because the largest body is the round form, it follows that it is the most spacious and judged by utility the most efficient shape for the earth. The third reason is necessity, because if the earth were not round, it would follow that there would have to exist either a vacuum or some body without space, which is impossible, and if the sky were flat, the space above our heads would be closer to us than other spaces and the stars there would be still closer to us than those of the east and west.

6. GRADUATE: I ask whether the sky has a color or what the color is which we see.[5]

COSMOGRAPHER: Our senses are often mistaken and our sight can be more easily deceived than any of the other senses; so, for instance, we see a whole stick put into water appear broken, and seen from a distance, two towers next to each other appear one. The sun, though many times larger than the earth, appears small; and it is known that above the moon nothing can be seen but the stars, which we see because of the light they receive from the sun. When we look up we meet no firm substance along the line of our vision upon which to fix our sight, as one might focus upon a wall or a similar object. Our vision grows weak, and as it cannot recognize an existing color, the eye produces its own color by its own humor so that the sky appears to have a color although the heavens as such have no color whatever except the pure clarity from the light of the sun.

7. GRADUATE: Of what size are the stars[6] and why do we see some with a tail of fire?

COSMOGRAPHER: As has been mentioned, the stars are in the eighth heaven which is their proper place; because of their great distance from us they appear small, although you should know that they are so large that if one of them were to fall the earth would be covered. The star which is seen at night running with a tail of fire is no star but fire,[7] as proved to the eye, and this fire originates in the air in the same manner as lightning. As will be shown, this is caused by clouds which make fire when they touch suddenly, or by vapors from the earth which are prone to these fires because of the sun's rays, and they disintegrate afterward because they do not contain lasting matter. Moreover, such fires are often caused in the daytime because of the blaze of the sun, but are not visible except when they have such force that their blaze overcomes the shining sun so that they are sometimes seen in daylight. These fire flashes serve as storm signals on land and sea when they occur frequently and in various parts of the sky.[8]

8. PILOT: Since in sailing the seas we see the altitude of the poles, I ask "What are the poles?"[9]

COSMOGRAPHER: The poles are two points which we imagine to exist in the sky. They are called poles of the world because they are the endpoints of the earth's axis, and so it must be noted that between the poles we imagine the whole of the round heavens to be divided into five zones. The first zone is that of the Arctic reaching from the pole to the Arctic Circle. The second goes from the Arctic Circle to the Tropic of Cancer, the third extends from the Tropic of Cancer to the Tropic of Capricorn and through the middle passes the equinoctial line. The fourth zone reaches from the Tropic of Capricorn to the Antarctic Circle. The fifth zone reaches from the Antarctic Circle to the Antarctic Pole. Thus the earth is proportionately divided into five parts by means of those five imagined zones, and it is to be noted that the circumference of the heavens contains 360°, so that we say that the distance from one pole to the other is 180°. It should be noted that those who live below the equinox and are in the middle of the surface of the earth have both poles on the horizon. People who live far from the equinox near one of the poles see but one pole, so that we see in the sky the point of one imaginary pole while the other stays hidden. Thus the one we see is called by the names Arctic Pole, Septentrional, and Boreal. It is called Arctic Pole because *artos* means *ursa major* [or "big bear"]; it is called Septentrional because there are seven stars near the pole. It is called Boreal because a wind called *boreas* comes to us out of the direction we call the north. The opposite pole is called the Antarctic, Austral, and Meridional. It is called Antarctic because it is placed almost opposite the Arctic Pole. It is called Austral because the wind that comes [from the south] is called *austral*. It is called Meridional because it refers to the position of the noonday sun with respect to our country.[10]

[*See figure, page 55*]

9. PILOT: Why it is that when we take the altitude of the North Star, we call "head" the part above, and "foot" the part below? How it is that there are head and foot in the sky?[11]

COSMOGRAPHER: Let me explain that there are four parts of the sky, east, west, south, and north. Those four parts we compare to the body of a man. We call the east, although most noble, the left arm;[12] and the west, the right arm; the north is called head and the south foot. Imagine a line drawn from the middle of the poles dividing the earth into two parts; the part which faces us we call head and the other foot, and by the same division we obtain the arms, to the east the left arm and to the west the right. Thus of the two parts created by the line one is called the upper part and the other the lower. It is advisable to know the region above the North Pole and the one below it; and of those two the part toward us, which as has been said is the part above us, we call the head and the part below us the foot. The reason for this is that people living below the equinox have the poles at the horizon, and as mentioned before, all the stars, those in the center of the sky as well as those close to the poles, rise and set for them each day. The North Star meantime turning around the pole gives a half turn over us, which limits the part called the head and is referred to as "above the pole" because it appears so to people living in the middle of the world; and the North Star will be, in the other half of its revolution, in the foot or "below the pole" when they cannot see it. That is the reason for calling what we see in the sky head or foot, meaning the head is the upper part and the foot the lower. Therefore, when the altitude of the North Star is taken and the star is below the pole, we add those degrees, and when it is above we subtract from the height the number as shown in the following figure.

[See figure, page 58]

10. PILOT: If the North Star goes around the pole, why is it sometimes 3° below and at other times less than half a degree?[13]

COSMOGRAPHER: Be it said that the North Star turns around the pole as all other stars do and that, in fact, in this turn the star does not approach, nor recede from the pole at any one time more than at another. But we say that sometimes it is more or fewer degrees above or below and this is why: When the North Star is in that region which we call west or east, it is neither higher nor lower than the pole but on a line with it. In its revolution from that position it rises or falls with respect to the pole, so that when it is in the northeast it is above the pole and when it is in the north—which is what we call the head—the star is as high as it can rise. When it is in the northwest it is already closer to the right of the pole, and in the west it is on the right of the pole. In the southwest it appears below the pole and in the south it sinks as low as possible. In the southeast it is below the pole. In this way, when the star is in the rhumb[14] of the east or west, it is neither higher nor lower than the pole, and in the northwest, north, and northeast it is above the pole. In the southeast, south, and southwest, it is below the pole. This explains how the star can be more or fewer degrees below or above the pole. The reason for subtracting from or adding to the degrees of the altitude of the North Star is to know how many degrees the pole is above our horizon; and that height is derived from the star and not from the pole because the pole cannot be seen, but the star in its circular course is in one-half of its path above the pole and in the other half below. So, when it is below, whatever degrees it lacks to the right of the pole are added to the degrees of altitude which we take; and when it is higher than the pole we subtract from the same altitude.

[See figure, page 61]

11. PILOT: Since the poles are not seen, how is it known in which rhumb the North Pole is with respect to the North Star?

COSMOGRAPHER: Although the poles of the earth are not easily seen, the rhumb of the North Star with respect to the Arctic Pole can be known by the rhumb in which the guards are, in this manner: It is said that the North Star and the guards turn around the pole. Since the North Star is nearer to the pole than the guards, its turning is less in circumference than the one the guards make. Thus, looking at the thirty-two winds for navigation [of a windrose], we find that the North Star is always twenty rhumbs above the rhumb of the guards. By this means, when the guards are in the northeast, the star will be with the pole in the southern quarter to the southwest. And if the guards should be in the north, the star will be with the pole in the southeast quarter to the south. If the guards should be in the northwest quarter to the west, the star will be in the east, neither higher nor lower than the pole. By these rhumbs, the rhumb of the North Star with respect to the North Pole can be known.

[See figure, page 63]

12. PILOT: Since the time at night is known by the guards, and if the North Star is so low that the guards (in part) are not seen, how do we know what time it is?

COSMOGRAPHER: In order to know and understand the hour at night, there are three other stars by which, even if the guards are invisible, time can be told in the following manner: the stars used are called "third," "sixth," and "ninth" and they have those names because the third comes three hours after the guard; the sixth, six hours; and the ninth, nine hours. So that if the guards are in the head, the third star is in the northeast, and the sixth in the east, and the ninth in the southeast, and in those directions one may find the others; so that knowing those stars—even if the guards are invisible—the time can be told by the course of each and also the point where the guards appear at midnight. Knowledge of those stars also helps in taking the altitude of the pole when the North Star is so low that the course of the guards cannot be seen, because through them the position of the guards is known.

[*See figure, page 65*]

13. PILOT: What are the Arctic and Antarctic Circles?

COSMOGRAPHER: The circles are at two imagined points, one on the circumference of the Arctic Pole, the other on the circumference of the Antarctic Pole; those circles are each 23° 33′ from the poles; and each of the circles is a like distance away from the two tropics of Cancer and Capricorn respectively, where the sun has a maximum declination of 90°. Thus from the Arctic Circle to the Tropic of Cancer directly [south] there are 90°, and in the other direction another 90° to the Tropic of Capricorn—we say the same about the circle of the Antarctic. People who live beneath those circles see the sun once a year for twenty-four hours continuously when it is on one tropic. But they will not see it at all for twenty-four hours when it is on the other tropic.

14. PILOT: As the tropics are 47° 6′ apart from one another, how is it that taken from one of the circles, each is in fact exactly 90° away?

COSMOGRAPHER: It is to be noted that wherever a man is, he sees half the sky and the other half is hidden and, as the sky has 360°, it follows that he sees 180°, that is 90° before him and another 90° behind him, and on each side 90°, so that he is in the middle of 180°. In each direction in which he looks he will see 90° of sky to the horizon. Thus I state that a man standing on the Arctic Circle can see each of the tropics at a 90° angle. When he faces north, the Tropic of Cancer at 90° distance is his horizon because 23° 33′ (which reach to the North Pole) added to 66° 27′ (from the pole to the tropic) make 90°, so that his horizon along that section is exactly the Tropic of Cancer. Turning his face south, he will see the Tropic of Capricorn at another 90°, of which 66° 27′ run from the circle to the equinoctial line, plus 23° 33′ from the line to the tropic, which adds up to 90°; so that from the same Arctic Circle his horizon goes to the Tropic of Cancer as well as to the Tropic of Capricorn. In this way, on whichever circle a man is, he is between the two tropics at exactly 90° distance.

15. PILOT: What is the horizon?[15]

COSMOGRAPHER: The horizon is a great circle around the earth on the surface or above the earth; it is called horizon, meaning the end of our view, because our view encompasses only the half of the sky which is above our hemisphere as distinguished from the other half over the hemisphere below.

16. PILOT: What is the equinox[16] and why is it found above the earth?

COSMOGRAPHER: The equinox is an imagined line along the middle of the earth, equidistant from the poles; it is called equinox because the sun crosses it twice during the year, that is, on the 11 of March and on the 13 of September.[17] The term equinox means equalizer, which refers to the fact that everywhere days and nights are equal, and from it the sun goes on to one of the tropics. It must be noted that those who live under the equinox live on the "right" sphere because both poles are at the same height; also one says "right" because their horizon divides the equinox at right angles. Those who live away from the equinox in our direction, which we call north, or the other, which we call south, we say have a "bent" sphere, one pole rising and the other setting because the horizon cuts the equinoctial line at unequal angles. The equinox is found during navigation of the seas by the altitude of the sun and of the North Pole; the number of degrees known by which the navigator is distant from the North or South Poles,[18] he can then set his course.

17. PILOT: What are the tropics and why are they so called?

COSMOGRAPHER: The tropics are two points or places which the sun reaches. One is its most northern point and the other its most southern. When the sun enters the first point in the sign of Cancer, and because of the movement of the firmament, it describes a circle which comes closest to the Arctic Pole.[19] This is called the summer solstice or summer tropic, and at this point the sun is closest to us. Another circle is made by the sun when it enters the first point of the sign of Capricorn, a circle which goes toward the Antarctic Pole to the south, and it is called circle of the winter solstice, where the sun is farthest from us. Each of the tropics is 23° 33′ away from the equinox so that from one to the other sphere there are 47° 6′, which is the latitude of the third zone, called by the ancients torrid, which means parched, because in that zone there is always sun and its movement is within the tropic circles. The word "tropic" derives from the greek *tropos*, which means wheeling, because as the sun goes through its zodiac on the way to the tropics, it wheels about and turns back.

18. PILOT: What is the zodiac?

COSMOGRAPHER: The zodiac is a circle in the firmament which cuts through the equinox and likewise is cut by it into two equal parts; one part leans toward the north and the other toward the south. It is divided into twelve equal parts. Each of these parts is called by a sign which has a special name of an animal or of another thing that agrees with it in likeness or property. These twelve signs of the zodiac are related to the behavior of the sun in each of those parts called a sign. Each one of these signs has 30° of longitude and 12° of latitude. Therefore, the zodiac has 360° in longitude. This zodiac corresponds to a band above the earth's surface which is 360° in circumference and 12° in width. The part of the zodiac which points away from the equinox toward the north is called northern [septentrional], and the six signs which are from the beginning of Aries to the end of Virgo are called septentrional [stars]. The other part of the zodiac which points away from the equinox toward the south is called southern [meridional], and those signs which are from the beginning of Libra to the end of Pisces are called meridionals or southern [stars].

19. GRADUATE: What is the sun, and why is it so called, and would its turning motion not be more useful than the enveloping one? Also whenever it crosses the equinox, why does the weather get milder?

COSMOGRAPHER: The sun [sol] is the fountain of light and is so called because it alone [solo] gives light to all living things which are alive because of it. If by its own motion it only moved from the west to the east in one year, it would go around the world for six continuous months above our hemisphere such that we would not have night, and for the other six months it would be in the hemisphere below such that we would not have day; thus it is very useful for the sun to cause the year's seasonal differences by moving around the earth's circumference in twenty-four hours. If the sun were always to go along the equinox, it would among other things damage the crops of the earth, which we cannot live without, nor would they ripen and we would not have the things which the sun produces for us as it does the different seasons we have during the year.

20. GRADUATE: How does the sun cause the seasons of the year?

COSMOGRAPHER: The sun marks the seasons of the year for us by approaching and receding from us in this way: when the sun is farthest from us it does not send out as much heat as when it is near; and as the earth and the water are by nature cold, the earth would get very cold because there is no heat, and with the cold the water thickens and clouds form, and when they disintegrate they cause rain. In this manner comes about the part of the year which we call winter, which is cold and wet. At that season the earth hardens through the cold and its accidental heat stays underground and nurses the roots of the trees. When the sun has come into Aries, which is in March, and it is neither very far from us nor very near, the weather is not very hot or very cold but mild, and so it creates the season we call spring. In this season the porous holes of the earth, which were closed in the winter cold, open up because of the sun's heat, and this passes to the roots and attracts the moisture. So it happens that the month of April [abril]—which means opening [abriente] because the earth opens in the said manner—is characterized by not having steady weather but, rather, rainy and clear periods and, likewise, now hot then cold ones. When the sun has reached Cancer, which occurs in June, when it is closer to us than at other times it dries the earth and constitutes the part of the year called summer, which is hot and dry. That season is characterized by the drying out of grass and tree leaves. In this way the sun creates the seasons of the year.

21. GRADUATE: If the diversity of the weather is due to the sun's being near to or far from us, when is the sun in Leo so that it is as far from us as from Gemini, in Virgo as from Taurus, in Libra as from Aries? Thus, why is the weather mild when the sun is in Gemini, Taurus, and Aries, and very hot when it is in Leo, Virgo, and Libra?

COSMOGRAPHER: We have explained that in spring the sun goes through the three signs of Aries, Taurus, and Gemini, which are in March, April, and May. During those months between the approaching hot weather and the cold of winter just past, the weather is mild. Moreover, when the sun passes into Leo and Virgo, that is, July and August, we get much hotter weather because the humidity of winter is spent by the heat of summer. Because that season is so hot and dry, the heat is almost gone when the sun descends into Libra in September and therefore the weather is cold and dry.

22. GRADUATE: Why is one summer or winter colder or drier than another or more or less humid?

COSMOGRAPHER: Let me explain that as the sun causes the four seasons of the year, the other planets also have their summers and their winters. It follows that if the sun causes our summer when another planet has its winter, the summer is then not so hot or dry as when another planet has its summer with the sun's and causes more heat and dryness than in another year. Similarly, when the sun makes our winter, and another planet has a summer, the winter will be less cold and humid for us; if the other planet has its winter with the sun at summer, it is colder and more humid. These reasons account for the rest of our weather conditions. It is also known that there are planets which are hot or cold or dry or humid, and therefore if in winter the sun is accompanied by a hot or dry planet, the winter will be less wet or less humid; if in summer the sun coincides with a cold and humid planet, it will be less hot and less dry in summer.

23. GRADUATE: Why does the sun seem larger to us when it is in the east or west than it does in the middle of the sky, since in the middle it is closer to us than in the other parts?

COSMOGRAPHER: This is because the middle of the sky is closer to us than the other parts, because between us and the center of the sky there is only air and fire; but between us and the east or west there is, in addition to those two elements, one-half of the earth, so that the sky above us is "closer" to us than the other parts. Moreover, the reason for the appearance of the sun is the vapors which rise between our eyes and the sun, and as the vapors are translucent bodies, they break the light beams traveling to our eyes in such a way that things are not seen in their proper size. This is what happens when a coin is thrown into clear water and, because of the deflection of the rays, it does not seem to have its proper size. That the sun is thus more distant from us in one part of the sky than in another does not contradict the statement of the round shape of the sky, because the sky is round and equally far from the center of the earth and we are on the surface or circumference of it.

24. GRADUATE: I have read that many suns and moons are seen together. I ask if this can be or whether it is an optical illusion.

COSMOGRAPHER: Leaving aside what we know by miracle [revelation] about what was made, in a more natural way [of knowing] I say that if several water pools are near us, we see in each one the image of the sun or the moon, and likewise we can see the image of the sun or moon in a cloud which is nothing but water. The Greeks call this phenomenon *perihelion* and it resembles the face of the sun only in size but not in light intensity nor in heat. There can be as many images as there are clouds opposite the sun or the moon to receive the image.

25. PILOT: When we sail on the ocean we direct ourselves by the altitude of the sun and that altitude is taken by shadows. This leads to the question, how many shadows has the sun on the earth?[20]

COSMOGRAPHER: The principal shadows which the sun makes of those who live on the earth are five in number, to wit: the shadow to the east, the shadow to the west, to the north, and to the south, and straight down. We say shadow to the east when the sun is in the west so that our shadow falls to the east. And we say shadow to the west when the sun is east of us and our shadow is thus to the west. We say also that the shadow is to the north when the sun is to the south of us and our shadow goes straight to the north, and the shadow to the south for those who live in the south is when the sun goes more to the north. There is also a vertical shadow when the sun is directly overhead, and the people living in the tropics experience all five of these shadows during the year. Thus the shadows are to the east when the sun sets, and to the west when the sun rises. Likewise, the shadows are to the north and to the south, and vertically because the sun passes overhead twice a year and then the shadow does not incline in any direction. People who live below the tropics have four shadows: to the east and to the west, and those of the Tropic of Cancer have a shadow to the north and those of the Tropic of Capricorn to the south; once a year they cast a vertical shadow when the sun is in either one of the tropics. We who live outside of the tropics have three shadows, that is, to the east and west and those who are in the north have a shadow toward the north and those of the south have a shadow to the south, and we never have a vertical shadow.

26. PILOT: I ask whether there is a place where there is no shadow at all although the sun is visible.

COSMOGRAPHER: I say that there is such a place where though the sun is seen, it makes no shadow to the east or west or north or south or straight down. This is proved thus: If someone were precisely beneath the Arctic Pole he would have neither an eastern [*levante*] nor western [*poniente*] direction because the sun does not rise [*se levanta*] whence comes the word for east [*levante*] nor does it set [*se pone*], whence comes the word for west [*poniente*]. Not having those two directions, a man has no north, being directly on the pole, nor south, for it is directly under his feet. So it is that he would see the sun for six months of the year, and even then he would never have any shadow from the above-mentioned directions because the sun could never at any time leave his zenith.

27. PILOT: Can one be in another place as far away from the sun as from the North Pole?

COSMOGRAPHER: That is quite possible at a certain time and place in the following manner: If you are in Seville on the 21 of October you are as far away from the sun as from the North Pole because Seville is at a point 38° from the equinox to the Arctic Pole and from the town itself to the pole, there are 52°. Now, on that day, the sun is 14° from the equinox in the direction of the Antarctic Pole, so that from Seville to the thirty-eighth parallel and from the sun to the fourteenth parallel there are 52°. Therefore, in this place and at this time, the sun and the pole are equidistant, and the same can be found to be true of other positions by checking the declination of the sun to the north and south parallels.

28. GRADUATE: What is the reason for an eclipse of the sun, and why is it larger at one time than at another?

COSMOGRAPHER: When the moon is in the head or tail of the Dragon,[21] or near it, and this happens in conjunction with the sun, then the body of the moon intercedes between our eyes and the body of the sun, and as the moon would be dark by itself, put as it is between the sun and us, we lose the light of the sun, and the sun appears in eclipse not because it lacks light but because we miss its light due to the moon's interposition between our sight and the sun. This eclipse always occurs in the new moon, which fact appears to explain that the eclipse during the Passion of our Lord, which took place during a full moon was not a natural eclipse but a miraculous one, because, as mentioned, the natural eclipse happens when the moon is new, and not at any other time.[22]

29. GRADUATE: Since we speak so much about time, I ask what is time?

COSMOGRAPHER: Time is the slow movement of the celestial bodies, which lasts as long as the movement of the heavens; from this it can be deduced that after the Last Judgment when celestial movement will stop, there will be no more time nor any difference in time. It must be noted that time was not created in time, because if it were created in time, it would mean that the time in which it was created was itself made, and it again made, in an infinite regression. Say that time is identified in five main divisions which are the year, the month, the week, the day, and the hour.

30. GRADUATE: What is a year and why is it so called?

COSMOGRAPHER: A year is the total time during which the sun passes through the twelve signs of the zodiac and returns to its starting-point. This it does in 365 days and 6 hours and this time is called a year [*año*], that is, zero [*anulo*], which is the same as a circle because the sun returns to its beginning. Before they made use of script, the Egyptians represented the year as a dragon biting its tail; but after they began to reckon the year, they started it in September when the trees bear fruit. The Arabs followed the same practice. The Jews begin the year in March, in which month they were given the laws and which they call the legal year. We begin our year in January, because the sun at that time begins to turn toward us.

31. GRADUATE: What is a month and whence does the name come?

COSMOGRAPHER: A month [*mes*] is a measure [*mensura*] or standard [*medida*] which measures [*mide*] the year, and it comes from *menne*, which is the Greek word for moon, and from that we have *month*. This includes the interval from the moon's departure from the sun to its approach to complete its circle. The Romans gave the months names for the gods they honored, and we follow their nomenclature. The Jews counted a month by the moon, calling the first day of the moon the first of the month and so on, so that however many days the moon had, so many had the month. Other nations have different and special names for months.

32. GRADUATE: What is a week and where does the word come from?

COSMOGRAPHER: A week [*semana*] is from the number seven [*siete*] since it contains seven natural days, during six of which God created all things, while he rested on the seventh which thus was sanctified. These seven days of the week take their names from the planets; as the sun, their king, reigns in the first hour of Sunday [*domingo*], that day is called *dominica*. Because the moon rules the first hour of Monday, that day's name is thus derived, and so on for the others. The Catholic church customarily counts the days of the week by market days, as second market day for Monday, third for Tuesday and so on except for Saturday and Sunday.[23] The Jews call them by numbers, as "Sabbath and one" for Sunday, "Sabbath and two" for Monday, and all the others following.

33. GRADUATE: What is a day and how many kinds of days are there?

COSMOGRAPHER: Day means light or brightness which the sun brings to the earth, and the word day [*dia*] refers to the gods [*dioses*] for whom the days were named; or they were so called from the Greek, which in Latin means double [*duo*] because the day is made up of night and light, or it is derived from *adian*, meaning brightness or light. Further, the day is to be understood in two ways: the natural day of twenty-four hours which includes the night and day, and the artificial day, which is the time when the sun is upon our hemisphere. The natural day has four qualities which are:[24] from the ninth part of the night to the third part of the day, hot and dry; and from the ninth part to the third part of the night, cold and dry; and from the third part of the night to the ninth part, cold and wet. So the artificial day when the sun shines for us has four parts: In the first, the sun appears reddish; in the second, it is bright; in the third, it is hot; and in the fourth, it sets and cools. For this reason they give the sun four horses according to the four different stages through which it passes. The Egyptians count the day from sunset to the next day at the same time. The Greeks and Persians begin with the morning; the Romans start the day at midnight, and each day goes to the next at the same time. Astrologers, Athenians, and Arabs start from midday; the Catholic church, in order to celebrate the holy days, chooses the beginning of vespers (evening); when the purpose is abstinence after the evening meal, the days begin at midnight, and the same holds for the observance and celebration of festivals with regard to the cessation of work.

34. GRADUATE: What is the hour?

COSMOGRAPHER: The hour is a length of time during which the sun passes half a sign of the zodiac. There are twelve signs of the zodiac; the sun passes all twelve in a natural day, which is the reason for the twenty-four hours. And because the sun at the equinox divides the signs of the zodiac equally, there are twelve hours in the day and twelve at night.

35. GRADUATE: Why are some days long and others short? What is the reason they lengthen and shorten?

COSMOGRAPHER: It has already been said that the artificial day is due to the light of the sun, and when the sun is nearest to us, the day is longer and the night shorter. In contrast, when the sun is away from us, the day is shorter and the night longer. Therefore, those who live directly below the equinox, as mentioned above, where the sun divides the signs of the zodiac equally, have days and nights of equal length, because they can always see six signs of the zodiac above their horizon. Those people have two summers because the sun passes the zenith overhead twice during the year, at the beginning of Aries and of Libra. They also have two winters, which come when the sun is in the Tropics of Cancer and Capricorn, because then the sun is farthest away from them. Moreover, for those who live in the north, the sun begins to rise from the first point on Capricorn toward Cancer; and as long as it rises, the days grow longer. For those in the Southern Hemisphere, the days diminish as the sun reaches Aries on the 11 of March the equinox, which is to say, that everywhere days and nights are of equal length. Once past the first point of Aries, our days begin to be longer than our nights, and in the other direction shorter. When the sun reaches the Tropic of Cancer on 11 June, we have the longest day and the shortest night; and the opposite is true in the other hemisphere because the sun is then closest to us, whence it turns and descends toward Capricorn. While descending it goes away from us and our days become shorter and our nights longer. When the sun reaches Libra on the 13 of September, it marks the equinox by making the days and nights of equal length; from there it begins to descend toward Capricorn and the nights grow longer than the days. When the sun reaches the tropic on the 13 of December, it gives us the longest night and the shortest day. To the people in the Southern Hemisphere, in contrast, the sun is closest when it is farthest from us.

36. GRADUATE: If when the sun approaches us we have a longer day, why is it that even though the sun is leaving the north, the people living closer to the North Pole and farther away from the sun, have a longer day?

COSMOGRAPHER: Be it known that those who live between the equinox and the poles have longer days the higher the pole is above their horizon, and this is why: Those whose zenith is on the Arctic Circle and for whom the pole rises above the horizon at 66° 27′ when the sun is on the first point of Cancer have a day of twenty-four hours and a bare instant of night, because in one moment the sun touches their horizon and then leaves, and that instant is all there is of the night. It happens, on the contrary, that when the sun is in the first point on Capricorn it produces a night of twenty-four hours and only a moment of day, because for an instant it appears on their horizon only to hide again, and that instant is all they have of day. In contrast, those who live below the Antarctic Circle whose zenith is between the circle and the terrestrial pole, experience a continued day without night while the sun is in their part of the horizon that cuts across the zodiac. If this were the extent of a sign, there would be continued day for a month, and in two signs it would be day for two months and so on. For those who live beneath the poles, all the year would be one day and one night, as follows: those who live beneath the Arctic Pole for the six months as the sun goes from the beginning of Aries to the end of Virgo—which are the six septentrional signs—would have a single day without night; and in contrast, when the sun goes from the beginning of Libra to the end of Pisces—which are the six austral signs—they would have one continuous night without day. Thus half the year would be day and half night. The same is understood for those living under the Antarctic Pole. And the reason for this is the global shape of the earth, which shrinks the horizon toward the poles so that those nearer the pole see more of the circle which the sun makes in the sky when it is in those parts. Thus the earth does not interfere with their seeing the sun, which they have before their eyes all the time that it takes to rise and to set until it reaches their horizon, where they cannot see it nor its return, and so the longer the time of its return, the longer the day.

37. GRADUATE: If from beneath the pole the sun is seen for six months continuously, I ask why there is less light and less heat?

COSMOGRAPHER: Be it said that our horizon has 180°, which is half the sky. The closer one lives to the pole, the more one can see of the course which the sun makes around the earth (when it is above that part), so that those who live on the circles see the sun's path completely on the day it enters the tropic, because their horizon reveals all the parts of the sky through which the sun turns on that day. For the same reason, those under the poles would see it for six months because their horizon is the equinox. Moreover, although all that time the sun gives light, it is oblique so that the nearest it gets to the pole is 66° 27′ away, so that although the sun is seen for six continuous months, another six elapse when the sun is not seen. Because of the great distance which it is from there, the sun gives less heat and light than in other parts since the continued cold raises the level of humidity to more than the sun is able to absorb.

38. GRADUATE: Since in some parts of the world the days are long and in others short, I ask whether during the whole year the sun can be seen for a longer time in one place than in another.

COSMOGRAPHER: Although the days are longer in one part than in another, as has been explained, it must be noted that the sun in its annual course around the earth is seen for the same amount of time everywhere—considering only the amount of day and night in each part. To wit: those who live under the equinox have nights and days of equal length, and therefore they have day half the time and night the other half; and those who have fifteen hours of daylight will have days of nine hours, because as the twelve-hour day increases to fifteen, so it diminishes from twelve to nine and the same happens with the nights. Those who have a twenty-hour day also have a four-hour day, and the nights are of equal duration; those who have daylight for a month without night also experience a month without day, and in this manner the rest can be computed. Explained in this fashion, it is clear that what in one part appears as day, in another appears as night. At the same place, therefore, there are equal parts of day and night during the course of the year.

39. GRADUATE: As the moonlight is caused by the sun, and the sun is always light, why does the light of the moon wax and wane?

COSMOGRAPHER: Some have said that the moon is itself luminous and that when it is in the same sign [of the zodiac] with the sun, the brightness of the sunlight would dim it, and at greater distance from the sun its light would become more apparent. This is not so, because the moon has no brilliance or luminosity of its own. Moreover, the sun, which is above it, lights it up every single day on our side; but when it is nearer the sun than we, the sun lights up the part above and makes a shadow on the earth and for that reason, we do not see it; that is, the moon is in conjunction [with the sun] and, as in its movement away from the sun it begins to shine a little and appears in the shape of a thin horn, the Greeks called it *mono dies* which means of one day, and as it gets farther from the sun and brighter it is called after eight days *dia thomos*, which means divided in half. As its light goes down, its shadow rises. After fifteen days it is called *anphitricos*, which means full, because it is farthest away from the sun. This is proved because when the sun hides from us in the west, the moon begins to rise in the east, and then we see its complete shadow rising and its light descending. After that it begins to approach the sun in the same way in which it had gone away from it. When the light rises, the shadow descends, and so it wanes in the same manner it had waxed.

40. GRADUATE: As the moon passes the zodiac in twenty-seven days and eight hours, why do we say that each moon has twenty-nine and a half days?

COSMOGRAPHER: This happens because in twenty-seven days and eight hours the moon traverses the zodiac. During those days and hours it does not overtake the sun, but when it advances for another two days and four hours it overtakes the sun and we have a conjunction of the moon and the sun. So we say that the moon has twenty-nine and a half days, understood as from one conjunction to another. Those who compute the moons account them twenty-nine or thirty days because they are not accustomed to count less than a full day. They begin their count in September in honor of the Egyptians who, as was mentioned, began their year then.

41. GRADUATE: What are the spots which we see on the moon?

COSMOGRAPHER: The spots which we see on the moon are produced in the following way: Let it be explained that although the lunar body is naturally dark of itself, as mentioned, and has no light at all, it happens that in some parts it receives more light and in others the light is less. The part which is more immediately touched by sunlight appears brighter and the part less so appears more turbid or darker, and this is how the spots which we see come about.

42. GRADUATE: What forms the circle which we see around the moon?

COSMOGRAPHER: As the moon is lower and older than any other planet and its shape is round, a circle is made when the air is neither very dark nor very bright, and the moon touches the air with its rays and makes a round figure in it; that circle is not far from the earth, but our own vision is deceived so that the circle appears close to the moon. Moreover, the air above is so thin that nothing could be made there, because forms must be made of thick bodies and the circle is less strong when the wind blows in midday. When the circle is of an even consistency below the moon and vanishes, this shows that the air is calm; and when some part breaks, it shows that the wind comes from that direction. Sailors take it as a sign of storm or strong wind. Such a circle also occurs in the daytime but cannot be seen because of the sunlight.

43. GRADUATE: It is said that the moon is in the head or tail of the Dragon. What does "Dragon" mean?

COSMOGRAPHER: All planets except the sun have three circles: deferent, epicycle, and equant.[25] The equant of the moon is a concentric circle with the earth in the plane of the ecliptic. The deferent is the eccentric circle, which is not in the plane of the ecliptic but rather one-half of it declines toward the north and the other toward the south and the deferent crosses the equant in two places. The outline of that figure is called "Dragon" because it is wide in the middle and narrow at the end. That is, the segment through which the moon moves from the south to the north is called the head of the Dragon, and the other segment, where it moves from north to south, is called tail of the Dragon.

44. GRADUATE: What causes the eclipse of the moon and why is it greater in one position than in another?

COSMOGRAPHER: As the sun is greater than the earth, half the roundness of the earth must always be lit up by the sun; the shadow of the earth is cast upon the air and diminishes in circumference until it fades in the plane of the zodiac, which is inseparable from the nadir of the sun. Nadir is the point directly opposite the sun in the firmament. From this it follows that during a full moon, when the moon is in the head or tail of the Dragon at the nadir of the sun, or near it, and the sun is diametrically opposed, then the earth interposes between sun and moon; the shadow of the earth falls on the body of the moon. From which it follows that as the moon has no proper light or brightness except from the sun, the light is blotted out and an eclipse results, which is bigger or smaller depending upon the shadow. This always happens at full moon or near it; however, at a full moon when the moon is neither in the head nor in the tail of the dragon, below the nadir of the sun, this does not happen.

45. GRADUATE: I ask whether the two outer elements [fire and earth] can exist without the others between or with only one between them.

COSMOGRAPHER: I say that the two outer elements, which are fire and earth, could never exist alone nor even with one other element between, which is proved in this manner. If there were no other elements between, there would be a vacuum between fire and earth; and as there can be no vacuum, the fire would descend to the earth or the earth would rise toward the fire. If earth would rise, while its nature is to fall, or if fire would descend, while its nature is to rise, the whole order of the world would be destroyed. Also, if those two elements of earth and fire were to touch with nothing between, the fire would enter the openings of the earth to burn and turn it to ashes; so neither men nor animals nor any other thing would exist. If only air were between, since it shares the nature of fire rather than earth, it would change into fire; if only water were between, since it shares its nature more nearly with earth than fire, it would change into earth, and thus earth would be drawn to fire and fire to earth, as has been said: so that because of the qualities of those four elements, the two inner must be between the two outer ones.

46. GRADUATE: What are the qualities of the elements, and could there be another substance between them which is not an element?

COSMOGRAPHER: The qualities of the four elements are that fire is hot and dry, air is hot and humid, water cold and wet, and earth cold and dry. They agree in weight, so that earth is heavier than water, which is heavier than air, and air weighs more than fire. As for something else between them other than an element, I say that there is no limit to the power of God, so that if he had made an outer one and had not made the air which man breathes and without which he could not live, nor without water which he needs continuously, there would be no men. So it is advantageous and necessary that those two intermediates be between the outer ones to temper all four elements.

47. GRADUATE: I ask if all those elements move, and why we see some of them and not others.

COSMOGRAPHER: Of the four elements the three which move are fire, air, and water. Furthermore, the earth is placed at the middle of the cosmos as its center, equidistant from the movement of all the spheres, and by its gravity and weight it remains motionless. Of the four elements we can see the two heavy and massive ones, which are earth and water. We see neither fire nor air because they are naturally light and thin and have no body; and we see only that which has body because our sight is coarse, and those two elements in their original state are so thin that sight cannot distinguish them.

48. GRADUATE: What is the element of fire and what kinds of fire are there?

COSMOGRAPHER: The element of fire is a very pure light which makes up the circle or the sphere which we call fire. It goes from the upper part of the air to the sphere of the moon, and although the actual fire which we see has three parts—light, flame, and red heat—it is true that the element is only light and not anything else, because the other two parts are material, or conserved in matter, but light is unique.

49. GRADUATE: What is caliginous air and why is it so called?

COSMOGRAPHER: The element of air, as has been said, is diaphanous and as bright as the light of the sun which it receives. By accident, however, one part may be brighter or darker than another, or hotter or colder. Be it known therefore that in this element or region of air there are three parts: one, the closest to us, then the middle, and finally the upper layer. The uppermost, because it infringes upon the element of fire, absorbs more of its light and heat and also the heat of the sun. The first part, which is nearest to us, heats up because of the reverberations which the sun's rays cause on earth. This it does in such a way that the middle part, which we call the middle region of the air, remains colder and darker because it does not share any heat and is by itself cold. Therefore, we call that part of the air "caliginous," or dark air, which because of its cold produces the rains, snows, hailstones, and other things we see.[26]

50. GRADUATE: What is wind and what is its cause?

COSMOGRAPHER: Wind is air which moves vigorously. Some say it is caused by the vapors of the water and earth which the air receives and that certain small things unite in the air, one pushing the other, and so they make the air move vigorously. They also say that the wind is made of humidity which stays in the earth in summer; later, when the sun sinks with much heat, during a great part of the night it draws the humidity from the earth and water, grows heavy, and so causes a vigorous movement. It is entirely certain that the wind is made up of great movements [of air] continuously carried by the waters from one place to another. And so we see that over the sea and near it, there are stronger and more continuous winds than on other parts of the earth.

51. GRADUATE: If the movement of the seas causes the winds, and since the seas move each day, why is there not a wind every day?

COSMOGRAPHER: All days have winds, although these are often not so strong that they reach us equally each day. Moreover many times, even if we do not feel winds, that does not mean that there are none in other parts. Also there are regions which have more continuous and special winds than others. So there is in Apulia a wind called "Arabolus" and in Calabria another called "Japix" and in France one called "Tirois" and in Vandalia or Andalusia another called "Solano," and so on in other provinces and places which have winds more constant or stronger than others, because they are close to a strait of the sea or near such places.

52. GRADUATE: If the wind is air and the air is naturally hot and humid, why is the north wind cold and dry?

COSMOGRAPHER: Nothing turns so readily from one condition into another as air, because it is between a hot element and a cold one and by nature it turns into their state. Therefore, when it is close to the earth it partakes of the qualities of the earth, and they are those of the place whence the wind comes. Such is the nature of the wind. So when the wind comes from a part or region of the east where it is hot and humid, the wind is hot and humid, and from the western region, which is cold and dry, the wind arrives cold and dry; and the two terminals of the earth, that is, north and south, are cold and humid, and for that reason the winds which come from them are cold and humid. Furthermore, although the north wind originates where it is humid, when it reaches us it makes a clear wind, which we call cold and dry; although the south wind originates where it is cold and wet, by the time it reaches us via the torrid zone, which is very hot, it heats up; since it comes with clouds and rain to where we are, we call it hot and humid.

53. GRADUATE: What qualities does our region have?

COSMOGRAPHER: The region in which we are has different qualities because of several winds. Because the part close to the mountains toward east and west is open to the south, it is hot and dry and good in winter. In summer, on the contrary, it is very bad. That part which is open to the east and closed in other directions is hot and humid and good in fall, and under opposite conditions it is hot and dry, very bad in fall and good in spring. And what is said here of the parts of the earth can be proved by the windows of the house, of which the ones facing south are bad in summer and good in winter, and those facing north have the opposite effect. For that reason the ancients built doors to their houses on the north and south. They ate and slept in the rooms to the south in winter, and in those to the north in summer.

54. PILOT: The maritime chart has thirty-two names of winds. I ask whether these exist everywhere.

COSMOGRAPHER: In any place where a man is, there are those thirty-two names of winds which are used in navigation, because although the element [air] is only one, we give names to the winds depending on which direction they come from. Accordingly, on account of the world and its roundness, the different names are shown in this form[27]:

[See figure, page 115]

and those names are neither Greek nor Latin, but current because of their use by navigators. Furthermore, it must be noted that he who is directly under the Arctic Pole has neither east nor west nor any other directions of the winds mentioned. As no part of water or earth can be without wind, the wind which exists is given the name of "sur" [rise] because as has been said, the property of that element is to rise [*subir*]; and at the pole the wind comes from below and so, as the "sur" [south] is formed below the North Pole, it follows that the wind has to rise up from there and not from any other part.

55. PILOT: I ask whether it is possible to sail with a wind to beneath the North Pole and what longitude has the world in each degree of latitude?

COSMOGRAPHER: I say that one can sail on any of the winds indicated on the maritime chart, and by the reckoning which will here be given, save for inconveniences, one could sail to below the north in the following manner: All the winds except the ones from east or west are called winds of altitude because sailing with them we gain or lose distance from the pole, so by any of them given from east to north one could sail to the place indicated, keeping in mind that with each wind sailed on a rhumb, for each 100 leagues sailed one loses $75\frac{1}{2}$ leagues' distance from the straight way, and for the half way $37\frac{1}{2}$ leagues, and for the quarter-wind $18\frac{3}{4}$ leagues. On the basis of this reckoning, if one goes around the world, arriving at the same point [of latitude] north and south from which departure was made, it is seen how far one is from the North Pole. One would have to sail on a quarter- or half-wind according to the course, more or less following the reckoning which has been given.[28] As to knowing the longitude which the world has at each degree of latitude, be it known that as one approaches the North Pole, the circumference of the world diminishes because the world is a spherical body. When one reaches the poles, its circumference is smaller still, and though everywhere the circumference of the earth is 360°, it is understood that as they approach the poles, the degrees have less distance, as seen in this figure.[29]

[*See figure, p. 118*]

56. PILOT: How does a whirlwind start?

COSMOGRAPHER: By watching the waves of the ocean, one learns that when waves approach the edge or shore with nothing in their path and hit a rock or similar object, then because of the great thrust which they impart, they turn over other waves and the water swirls around and makes a whirlpool. By analogy, when the wind has no obstacle in its way, it blows straight, but when it is interrupted as by a mountain or other opposing wind in front, it turns around and whirls.

57. GRADUATE: What is the cause of rain and snow and hail?[30]

COSMOGRAPHER: Rains are caused by heavy vapors and humidity from the earth and water, which are attracted by the sun and rising up condense with the cold and form small drops. As the drops rise they enlarge. Also the air, due to the cold of the earth and water, thickens and changes in substance, which we then call clouds, which cause more rain the thicker they are. The snow is formed in that wet vapor which rises from the earth because of the cold which is close to the earth. It contracts and thickens and falls as snow. Hail is caused when the vapor of water and earth rises higher up; and because up there the air is very cold, it freezes the drops of water and turns them into stones, and as water drops are round, so hailstones are round. That is, the first condensation is of water, the second of snow, and the third of hailstones, and this we see happen at the end of spring when hailstones fall. In that season snow, which comes down in the winter, does not fall. The reason is that in summer vapors rise higher, because of the sun's heat, and while they rise the droplets freeze and become bigger and the ice drops form hailstones. In winter when the cold is close to the earth, the water contracts before thickening because of the cold of the season, and turns into snow.

58. GRADUATE: Why, in the high mountains which are nearer to the sun, is there snow all year round when there is none in the valleys?

COSMOGRAPHER: It has been said that air has three parts, which are the first, the middle, and the upper; and the higher it is the purer air gets, as well as thinner and colder in the process. When it is lower and close to the earth, it is denser and less cold, because as the rays of the sun touch the valleys and the sides of the mountains and cannot go on, they rebound there and heat the earth and for that reason the air close to the earth is less cold. Because of the cold of the air, there is more snow on the higher parts than in the valleys.

59. GRADUATE: How is a rainbow made, which appears in the sky when it rains, and what causes its colors?

COSMOGRAPHER: "Iris," or the rainbow which we see during rainy weather is caused by the reflection of the rays of the sun when they shine upon some watery cloud, and they are caused on the one hand by the sun and on the other by the cloud. As the rays of the sun touch the cloud, they break it and then appears the rainbow; where it is bright it takes on various colors. They also say that the rainbow is the image of the sun formed in the cloud, and as every image has the shape of the original thing it represents, and as the sun is round, the bow appears in a round shape. Since the bow has no substance, it must be said that by itself it is the image of substance and as such could have no colors but only images of colors. Others say that the rainbow is a thick luminous cloud which has the four principal colors of the four elements: the red of fire, the blue of air, the yellow of water, and the green of earth due to grasses and trees. Also they say that the mixture of the clouds with air and fire and the sun's rays result in that variety of colors which we see.

60. GRADUATE: If the rainbow is made in the image of the sun, and the sun is round, why do we not see a complete, round rainbow?

COSMOGRAPHER: As the sun is much higher than the clouds, when it touches a cloud from above it makes its image in it, and as we are below we cannot see it all. The closer the sun is to us from the east or west, the more we can see of the circumference of the rainbow.

61. GRADUATE: What causes thunder and lightning?

COSMOGRAPHER: Thunder results from the touching of currents of air which run into each other with great force, and lightning is air which has set itself on fire and lights up.

This comes about in the following way: the wet vapor, as explained, rises and as it approaches the upper air, the currents from one part clash against another, and from this collision derives the noise of thunder. This motion heats up the air to such an extent that it is converted into the substance of fire, which causes the explosion or lightning which we see when there is thunder. Although this happens all at once, we see the lightning sooner than we hear the thunder because sight is quicker than hearing. Also it is said that the collision of the watery cloud as it rises and meets its opposite, which is fire, causes thunder and lightning.

62. GRADUATE: What causes comets?

COSMOGRAPHER: Matter, from which derive all things made of elements, originates them in two ways: from a hot and dry state, called exhalation, and from a hot and wet one, called vapor. From the hot and dry state comets are made in this way: when an exhalation rises to the sphere of fire, it burns there, so that we see the fire [of a comet]. Because there are planets that attract some exhalations, it is said that these exhalations form the substance which is seen in the sky as a path of clouds, which is commonly called the Way of Santiago [The Milky Way].

63. GRADUATE: What are rays, and what effects have they?

COSMOGRAPHER: Rays originate in two ways and are of two kinds. One is part of the air, which turns into fire and rises, and this happens at the time of thunder and lightning, as was said, resulting from the movement of those parts of the atmosphere against the other principal part. This part rises and catches all the fire, which it disperses. The fire then descends and runs from one part to the other with great force until it forms as large a thing as what opposed it; and those rays burn the things which are high up and produce great effects, as was said before. There are other rays, of the substance of stone, originating in the following manner: When the humid vapor rises and has in it something made of the substance of earth, this condenses because of the heat of the sun and turns to stone, and is held in the cavity of the cloud until this opens up and then the stone descends and injures wherever it goes. It cracks towers and buildings and strikes ships and pierces the earth and does other great damage. We see the serious effects of these rays; in a particular case, a man dies from the ray's drying his bones without penetrating the skin. In jointed wood it destroys the nail while the wood stays whole. It melts bells, while the ropes do not burn; it breaks and burns a cask, but the wine does not spill for three days.

64. PILOT: What is the sea and why is it called ocean?

COSMOGRAPHER: The sea is a gathering of the waters, and is the same element as water. It is called ocean from *okys* in Greek, which means speedy. The Greeks and Romans called it by that name—ocean—because of the speed and acceleration with which the sea runs all the time. It is not uniform in its movements; for seven days it rises or swells, and for another seven it retreats or falls, and its color changes according to the variety of the winds.

65. PILOT: Why does the sea rise and fall each day?

COSMOGRAPHER: The wisdom of God saw that without heat and moisture nothing could live. Since the earth is cold and dry, He put the source of the sun above it. And because nothing can live only by the sun's heat, He put the land in the middle of the source of moisture, which is the sea, to moderate the sun's heat. Thus these two things moderate each other. The sea, which comes from the west has two parts, one toward the south and the other toward the north along both sides of the shore, and similarly, toward the east there are another two parts, so when the two branches of the sea from the west meet, the great impact reverses the seas, and some go and others come; so it is when the other two arms meet in the other part and have the same effect, causing the ebb and flow. Others say that the elevations and great depths which are in the sea make the sea turn from one part to another. The most certain fact is that the sea, which is the element of water, moves naturally like the other two elements of fire and air, and this movement comes so regularly with the waxing and waning of the moon that some say that the moon causes the high and low tides of the sea, because the high tides come exactly with the moon. So as the moon rises and sets each day at different hours, at the same hours the tides occur.

66. PILOT: Why is the sea salt?

COSMOGRAPHER: Most of the ocean water is below the torrid zone, which is the part where the sun is hottest; and because of the great heat of the sun, it sucks up the waters near it that are purest and lightest, and the heat of the sun consumes them, as do the attacks of the winds. What is heavy and earthy remains, and the sea evaporates and becomes bitter or salt. Since, as has been noted, water turns salty when the strong heat of the sun is on it, it is said that the sea is saltier in the fall than at other times because of the heat of the summer which has passed over it.

67. PILOT: Is the sea round or level, or what shape has it?

COSMOGRAPHER: The water of the sea is round and that is so for two reasons: first, as the water is a body of one shape or nature and all its parts are of the same shape, so every one of the parts has the shape of all, because by nature the parts of water want to have the form of the whole. We see that the waterdrops which fall as dew on the grass are round, and they are as parts which show the shape of the whole. The second proof is had by putting a marker on the shore of the sea, and when a vessel leaves and as it gains distance from the shore, a man standing by the foot of the mast, keeping the marker in sight will soon no longer be able to see it, but if he were to climb up to the top of the mast he would still be able to see the marker, although, in fact, standing by the foot of the mast he is closer to it than he is at the top of the mast; and this is due to the curvature of the earth.

68. PILOT: If a pilot on the ocean were to lose his chart and compass, how could he steer to make his voyage?

COSMOGRAPHER: Although in this case additional details of the event are necessary, I say that the pilot should take two things into account: first, his course followed according to his intended route, and second, the time when such a thing happened. Depending upon these, the sun will serve as compass and his knowledge of winds as a course in the following way: on the 11 of March, the sun rises in the east and sets in the west and one must know how between the 11 of March and the 11 of June—that is, during three months—the sun slows its rising from the equinox to the Tropic of Cancer. It departs from the said line by 23° and 33', which it does in this manner: the first month 11° 47', and the second 7° 52', the third 3° 55'. Dividing the 360° of the circumference of the earth by the thirty-two winds of navigation, there are from one wind to another 11° 15' so that if the sun is 23½° from east to north, it is two rhumbs away; so we say that if the sun is in the Tropic of Cancer, it rises ENE and sets WNW. And in the same way by which the sun rises from the line to the tropic during those three months, it recedes in another three, from the tropic to the line; and on the 13 of September it turns, to rise in the east and to set in the west. In this manner one can count off the other six months of the year when the sun goes to the southern part for three months, dropping from the line to the Tropic of Capricorn, and the other three months while it returns from the tropic to the line. So depending upon the time when the accident happens, the rhumb where the sun rises and sets can be known. Moreover, as there are twenty-four hours and eight winds in the natural day, there is a wind every three hours, half a wind in one and a half hours, and a quarter of a wind in three-quarters of an hour. Checking this reckoning by the clock one will know for certain at each hour the rhumb in which the sun is. And this is done thus: on the 11 of March the sun rises at six o'clock in the east and at nine o'clock it will be at southeast and at noon it will be south, at three o'clock in the afternoon at southwest, and at six o'clock it sets in the west. According to this rule, one can distribute the halves and quarters and so the course or way to be followed is known. Also the more hours the day has, the more winds the sun passes through by day and less by night; in contrast, when the night is longer, the number of winds which the sun goes by day are fewer, and more by night, so that if the day had twelve hours the sun goes four winds by day and four by night, and if the day has fifteen hours, it goes five winds by day and three by night; and in a day of nine hours, three winds by day and five by night. So can be counted the rhumbs which the sun goes in the day and at night depending upon the time and place where one is.

69. GRADUATE: How is the earth situated in the universe?

COSMOGRAPHER: The element of the earth is put in the middle of the universe because it is the lowest, and as the world is described as round, and everything in it is round, the middle is the lowest. That the earth is in the middle of the firmament or eighth sphere is proved, because to those of us who are on the surface or circumference of the earth, the stars appear of the same size in any part of the sky no matter whether they are to the east or to the west. The reason for that is that the earth is equidistant from the stars in all parts. It follows therefore that it is in the middle of the firmament, and if at some point the earth were closer than in another, a man in the part closer to the firmament would not see the middle of the sky, which is against Ptolemy and all the philosophers, who say that from any place in which a man is born he sees six signs, and has another six to see; one half of the sky is visible, and the other half is hidden. He is equidistant from the firmament, which is round, and he is the center or hub of that same firmament.

70. GRADUATE: As the elements are so close to each other, I ask if the earth is below or above the water.

COSMOGRAPHER: Any one of the three elements can fit closely around the earth except when the dryness of earth resists the humidity of water, in order to support the lives of men and animals who live by respiration. Some say that water does not cover all parts of the earth because of the influence of certain stars which are close to the Arctic Pole; and this seems to be so, because the Arctic Pole attracts dry things and the Antarctic Pole humid things. This is proved by the lodestone, of which one part indicates the Arctic Pole, to which it is attracted by iron, and the other correspondingly points away.

71. GRADUATE: Now, as the earth is in the middle of the air and is heavy, why does it not fall, since everything else in the air falls?

COSMOGRAPHER: Although the air surrounds the earth on all sides, there is no place for the earth to fall as there is nothing lower to hold. Because it is lowest it cannot descend, and nothing holds it, although some say that the earth is held in the air like a ship on the water. Others say that as fire envelops the earth with a force which would pull the earth up, it cannot move, neither up nor down nor in any direction,[31] as is also stated with reference to the coffin of Mohammed, which is of iron and locked by a lodestone. This means that the natural virtue of stones maintains the earth in the air, or, as the earth is in the middle of the element of fire, the fire's virtue sustains it. The true reason is that as all things weighty and heavy go to the center because the center is the fulcrum of the firmament, and because the earth has great weight and heaviness, it naturally finds itself at the center of the firmament and all that moves rises from the center. So if the earth moved from the middle or toward the circumference, it would have to rise, which is impossible, so the earth is in the middle and neither moves nor could move.

72. GRADUATE: I ask if the earth is round or flat, or what shape it has.

COSMOGRAPHER: The earth is round, although some believe, more due to sight than to reason, that it is flat. But this is not so, because if the earth were flat, the waters of the rain which fall on the earth and rivers would no longer run to join and form lakes or pools in one place. Also there are stars which appear in one zone and not in another. The Egyptians see a star called Canopus which we do not see, and that would not happen if the earth were flat; also it appears that the earth is round everywhere from the [zodiacal] signs and stars which would not rise where they do rise and set if they were seen equally by all men in any part [of the earth]. They rise and set earlier for some than for others, and the swelling or roundness of the earth causes this. So we see that an eclipse of the moon does not appear simultaneously to everyone. The roundness of the earth from north to south and in the opposite direction is also shown thus: Those of us who are toward the north always see some stars which are near the Arctic Pole, and other stars which are near the Antarctic Pole we can never see. Furthermore, if someone from the north were to walk south he could go so far that the stars which he first saw would be hidden from him and he would not see them; and the further south he went, the less he would see of the stars which are close to the north. Then he would see the stars of the south, which he at first did not see, which are near the Antarctic Pole, and the contrary would happen to him walking south to north, and the sole reason for this is the roundness of the earth.

73. GRADUATE: How can the earth be round, when there are so many deep valleys on it and mountains which reach to the clouds?

COSMOGRAPHER: Our small size is responsible for making small things appear large to us. Take the great height of that Mount Olympus and all the earth in comparison to the sky, it certainly is very small or nearly nothing. Take the case of ground passed over in one step, which might cover a small difference of elevation of the ground. It would certainly seem to us that there is neither valley nor mountain. A great valley would thus seem to be a small thing indeed with respect to the heavens. Therefore neither valleys nor mountains affect the roundness of the earth. It should also be noted that anything is called round for two reasons: one, a regular feature, the other an irregular one. The regular one is when lines drawn from the center to the circumference are equal, but this is not the case in the roundness of the earth; the other one is irregular roundness, when all the parts are not equidistant from the center, and that is the kind of roundness the earth has.

74. GRADUATE: What size has the earth as compared to the sky?

COSMOGRAPHER: The earth is so small with respect to the sky that it is almost unnoticeable, which can be demonstrated in this way: The smallest star of the firmament seen by the naked eye is larger than all the earth. One star—with respect to the sky—is almost nothing, because even in only that part of the sky which we see there are so many stars and, besides, there is much space without any stars. Thus, how much smaller is the earth than the sky, since the earth is smaller than a star. That the earth itself is small with respect to the sky is proved by the sun, which at night, being below our hemisphere, sends its light to all the stars above us without being hindered by the roundness of the earth and the water in the center. So it is proved just how small the earth is with respect to the sky.

75. GRADUATE: If the earth were bored from one part to another through the middle, and a stone were thrown into the hole, I ask where would it stop?

COSMOGRAPHER: Let it be said that the earth is lower than any other element, and for that reason the middle of the earth which is in the center is still lower than any other part. The stone which the man throws into that hole would fall to the middle of the earth where the center is, and naturally there it would stay because if it went farther it would have to rise, which as it has weight it naturally cannot do.

76. GRADUATE: How are fountains made in the earth and why are some waters sweet and others salt?

COSMOGRAPHER: Let me explain that the sea is the element of water, and from it come all the waters and to it they all return. It has already been said that the waters return to the place whence they came in order to run again. So when the water runs through the hollows of the earth, if at the end of the hollow there is something hard which it cannot pass over, nor can it turn back because of the force of the water behind, it sprays up over the earth and makes a great or a small fountain, depending on the quantity of water. And as water is smooth and passes and winds through the holes in the earth, it takes on various tastes, so that if it passes through a sandy or rocky place it takes on a sweet taste; and when it passes through salt earth it takes on a salt taste; when it passes through muddy land it comes out with a bad taste; and when it passes through stones of copper, chalk, or alum it is bitter; so that according to the diversity of the earth the water takes on different tastes.

77. GRADUATE: Why does water from the wells not swell or spill over like fountains?

COSMOGRAPHER: Although well water originates in the same manner as does fountain water, the well does not swell or spill over because where the water appears, there is no place for it to run and it does not bubble up. That well water comes through caverns in the earth is proved by the fact that near rivers there is always water, and if a well is dug close by another the water will run from one to the other. Although there are some wells which have no holes or caverns through which the water reaches them because they are in high or dry places and still have water, this is because the earth, though naturally dry, has by accident collected some humidity by distillation, which falls drop by drop into the well. That the water is collected by the perspiration of the earth is proved by the fact that there are water-filled wells in dry places.

78. GRADUATE: Since the earth and the water are naturally cold, why is well water warm in winter and cold in summer?

COSMOGRAPHER: It has been said that in the cold of winter the pores of the earth close up, and as the vapor cannot leave and remains inside, the water is warm. In summer, when the earth opens up in the heat, the vapors rise, and the heat is easily dispersed, the water comes up cold. The principal reason is that all things have their likeness and their opposites. In winter the air is cold on the earth, the heat flees to within and hides, and the well in winter is warm; in summer the water above is burnt by the heat of the sun, and the cold of the earth flees within and makes the water cold. In summer the water of the fountain is less cold than that of the well because it is nearer the air, and that of the river is still warmer for the same reason.

79. GRADUATE: It is said that there are men called antipodes who walk about on the other part of the earth below us. I ask how can they walk below us.

COSMOGRAPHER: As the earth is round, which has been proved, the people living on the other part of the earth opposite to us have their feet toward ours, as we do for instance when a man close by some lake sees his image with the head in the other direction and the feet against his feet. So, the people we call antipodes, which means they have their feet against ours, have their heads toward the sky as we do. Therefore it must not be thought that we are above them or they on top of us because a round body has properly no upper part, so between their feet and ours there is the round body of the earth and waters. Let it be further known that they and we do not have the same summer or winter, nor do we share the other seasons of the year. Rather, when we have summer they have winter, or the opposite, and when we have day they have night, because the sun does not shed light simultaneously upon them and us, and for that reason we have different seasons. As the light of the sun causes the day and the shadow of the earth the night, so when the sun is above in our hemisphere the shadow is in the lower hemisphere, and the contrary. In this way there is no hour when it is not night somewhere because of the roundness of the earth.

80. GRADUATE: What is a hemisphere?

COSMOGRAPHER: A hemisphere is one half of the sky: so there are two hemispheres, that is, the upper one and the lower one. The upper one is that half of the sky which our eye perceives as a rim or at the level of the earth, and we call it upper because with respect to us it is above the earth which we see, and the hemisphere below is that half of the sky which we do not see. The word hemisphere means one half of the sphere.

81. GRADUATE: What is climate? How many climates are there? What name and extent does each one have?

COSMOGRAPHER: Climate is a line drawn from east to west equally far from the equinox.[32] The ancients divided the world, known to them as inhabitable from the equinox to the Arctic Pole, into seven parts or climates, and they called them after the city or mountain or river or other famous place where they marked them; and those climates are unequal in longitude as well as latitude, because in a longitudinal direction the roundness of the earth contracts from the equinox to the poles, as was said, so the longitude of the first climate is larger than that of the second and so on for the others. The names and width of each climate are the following:[33] The first climate is called Diameroes after the isle of Miroc and the River Nile. It begins 12° 45′ from the equinox and goes to 20½° and its width is 7° 45′. The second climate is called Diasiene after Siena, a city in Egypt. It begins at 20½° and goes to 27½° and has a width of 7°. The third climate is called Dialexandros after Alexandria, the city in Africa. It begins at 27½° and goes to 33° 40′ and has a width of 6° 10′. The fourth climate is called Diarodos after the Isle of Rhodes. It begins at 33° 40′ and goes to 39° and has a width of 5° 20′. The fifth climate is called Diaromes after the city of Rome. It begins at 39° and goes to 43½° and has a width of 4½°. The sixth climate is called Diaborestenes from a river in Scythia. It begins at 43½° and goes to 47° 15′ and has a width of 3° 45′. The seventh climate is called Diarifeo after some mountains in Germany. It begins at 47° 15′ and goes to

$50\frac{1}{2}°$ and has a width of $3° 15'$. It should be noted that each of those degrees is $17\frac{1}{2}$ leagues long,[34] and by this measure the width of each climate is known, counting 60 minutes per degree. Although there are lands, islands, and dwelling places beyond them, the people there find themselves living in an intemperate climate never encountered by those below it. By experience we see that below the equinox, which region they [the ancients] regarded as uninhabitable, believing that there is great heat, there is in fact land with a very temperate climate, and the reason, as before said, is that as days and nights are equally long, the heat of the sun during the day is moderated by the cold of the nights; so all year round there is a continuous moderate climate. But in each of the different climates there are different days and nights.

82. PILOT: In navigation one goes by distance and direction: I ask what they are.[35]

COSMOGRAPHER: In navigation upon the sea three things are especially used. They are called wind, rhumb, and distance. The wind, as already described, has thirty-two different names, divided into three parts which makes eight principal winds and eight half-winds and sixteen quarter-winds. The eight principal winds are these:

[*See figure, page 152*]

These winds are called in the Italian language:

[*See figure, page 153, top*]

The other eight, which we called half-winds or halves, are the ones above described, which are so called because they are between the principal ones, and their names are these:

[*See figure, page 153, bottom*]

Between those sixteen winds are another sixteen which are called quarter-winds and which have the names of the collaterals, as pointed out in question 54. Rhumb is that line or curve by which those winds are given on nautical charts; these rhumbs are like characters (for each of the winds) and show the direction they came from and where they are going to. And as before said, there are half- and quarter-winds. They are indicated as black (full), green (half), and red (quarter) winds.[36] The dark ones show the eight principal winds, the green ones the halves, and the red ones the quarters. By the means of these rhumbs, the winds are distinguished and identified. A course is the way traversed on the sea under sail, which is given by one wind together with another, so that we describe the course in this way: He who is in Toledo and wants to go to Seville would be on a course N–S, which means leaving the north at the back and facing the south, or noon. But if he went from Toledo to Granada his course would be NW–SE, which is turning the face to the south-east and leaving the north on the left side, so that where people are with respect to the wind describes a course in navigation.

[See figure, page 155]

Notes on the Translation

1. The Greek and Latin terms in italic are transcribed, not translated or put into modern transcription, because Medina intended the derivations to serve as memory devices for sailors.

2. For the history of this concept, see Thomas S. Kuhn, *The Copernican Revolution*, chap. 1, "The Ancient Two-Sphere Universe."

3. The count of spheres differs in this book from that in any other by Pedro de Medina. He omits the crystalline as a separate sphere to be drawn into his diagram, because the so-called second-mover is a construction of physics, not of astronomy. The crystalline sphere was imagined as necessary to cool the friction resulting from the "abrupt" movement of the primum mobile. In all later books, Medina draws the customary eleven spheres. On the eleven spheres in Medina, see Salvador García Franco, *Historia del arte y ciencia de navegar*, 1:141–43.

4. D. E. Gershenson and D. A. Greenberg in "How Old Is Science?" (*Columbia University Forum* 7 [1964]: 26) state that "analogy is one of the most powerful and ubiquitous tools of human thought" and give examples from Aristotle and Anaxagoras of ancient scientific thought and method which are of interest in this context. See question 57 of this text and note 30. See also G. E. R. Lloyd, *Polarity and Analogy: Two Types of Argumentation in Greek Thought to Aristotle* (Cambridge, 1966).

5. This question did not get any better answer for some time. See, for instance, Isaac Cardoso, *Philosophia libera in septem libros distributa in quibus omnia quae ad philosophum naturalem spectant* (Venice, 1673), bk. 2. It received reconsideration in the light of modern knowledge by the Hon. John W. Strutt (later the third Lord Rayleigh) in his famous essay, "On the Light from the Sky, its Polarization and Colour," *Philosophical Magazine*, 4th ser. 41 (February 1871): 107–20.

6. For the size of the stars, in his other works Medina relied upon a book on the spheres by the ninth-century astronomer al-Farghani.

7. Comets were not regarded as celestial bodies before Copernicus. See Kuhn, *The Copernican Revolution*, p. 45.

8. This passage can be found in almost all contemporary cosmographies. The flashes are usually called Saint Elmo's fire.

9. This question, the first addressed to the cosmographer by the pilot, introduces issues relevant to the art of navigation (see above, chap. 3).

10. This answer explains the role of the celestial pole in obtaining latitude. The angular distance between the pole and the horizon equals that between the equator and the point of observation as measured from the center of the earth. Ascertaining the position of the celestial pole was therefore the first task of the navigator.

11. Fifteenth-century astronomers knew that the pole-star did not indicate the true elevation of the pole but was distant about $3\frac{1}{2}°$, making a full revolution around the pole about every twenty-four hours. The navigator needed to find out where the celestial pole was with respect to the polestar. This he could ascertain by observing the polestar "in rule," i.e., with regard to the position of the guards of the Little Bear in their revolution about the polestar; and from that datum it was known whether the polestar was to the right or left or above or below the celestial pole. The correction for each hour of the night was given in the "Regiment of the North," a table submitted by Medina separately along with the *Libro* in 1538 and preserved in its earliest version in the *Coloquio* of 1543. The table was published in very slightly corrected form in the *Arte* and *Regimiento* of 1552. The sky-man of this question, or clock in the sky, was an image in use among experi-enced sailors when Ramón Lull made it famous in his dialogue of 1286: *Finis de las maravillas del orbe*. The midnight position of the guards of the Little Bear shift about an hour every two weeks and this position could be memorized at least for each month. The man in the sky with the polestar in his breast, his head to the north celestial pole, etc., permitted telling time like this: "mid-July, midnight in the right arm; end of July, one hour below the right arm,' etc. (Taylor, *The Haven-Finding Art*, p. 146). The figure of the sky-man had been amplified by adding four lines midway between his arms and legs so that a correction could be made for each of eight points of a circle (divided into "winds" or directions), which is the way Medina makes his illustration. From the sky-clock was obtained the number of degrees to be added to or subtracted from the observed height of the polestar in order to obtain the angular distance of the celestial pole above or below the equator, which is the latitude.

12. Eastward toward Jerusalem. Right is more noble than left in the Christian tradition; left is more noble than right among Moslems. Taylor in "The South-Pointing Needle," pp. 1–8, indicates various orientations. The sky-man faces the observer (west = right).

13. Waters, *Art of Navigation in England*, p. 46, fig. 4.

14. See question 68 and answer. Rhumbs of the winds are the thirty-two directions of the seaman's horizon given on a sea chart and compass rose. See Waters, *Art of Navigation in England*, pp. 48–49, on finding latitude by Polaris.

15. The apparent horizon.

16. Also called the celestial equator.

17. These dates are all from the old calendar in force at Medina's time. In 1582 the Gregorian reform put the

calendar ten days ahead when Thursday, 4 October was succeeded by Friday, 15 October. The dates in this translation are not corrected.

18. That is along the latitude.

19. The sun is retreating eastward.

20. Determination of latitude by the sun involved more extensive calculation and data then by the north celestial pole. The sun's angular position north or south of the equator could serve in the same way as the altitude of the pole. Using the sun required observing its meridian passage, which is the highest point above the horizon. The number obtained had to be recorded, together with a number giving the declination of the sun for the date; then, according to certain rules, these had to be converted into the angular position sought. This calculation required a knowledge of the passage of equinoxes and depended on whether the pilot was in the northern or southern hemisphere. In fact the navigator had to convert the observed altitude of the sun into terms of the celestial equator's altitude, which was then added or subtracted from 90°. This was a matter of eight steps (Taylor, *The Haven-Finding Art*, p. 165). The observation was made with either the cross-staff (*ballelstilla*) or the astrolabe. The use of both of these instruments is shown in Medina's *Arte* and *Regimiento*s, and they appear as illustrations in his *Suma de cosmographía*s. Preference depended somewhat upon latitude. The cross-staff was very difficult to use at or near the equator (L. C. Wroth, *The Way of a Ship*, pp. 25-28).

In his "Regiment of the Sun" Medina offered the pilot a conversion of the sun's observed meridian passage (diagonal movement) relative to the celestial equator along the ecliptic into its vertical movement north and south of the equator. The calendar of the sun's declination was accurate only over a period of about twenty years, and account had to be taken of the four-year cycle of leap years, so that there was much

occasion for presenting better or more recent tables. The prolix treatment of shadows in the observation of the altitude of the sun is a common feature in Spanish cosmographies of the time. Kuhn (*The Copernican Revolution*, p. 10) gives an illustration of shadows. Waters explains in a note that the rules which this text represents are now given by "the simple formula, $l = d - Zm$, where the latitude, l, is expressed as a function of the sun's declination, d, and of the meridional zenith distance, Zm, northern declination and latitude being considered as positive ($+ve$), southern as negative ($-ve$), and Zm as positive or negative according to whether observation was made towards the north or south" (*Art of Navigation in England*, pp. 50, n. 1, and 51, fig. 6). Medina's earliest tables of the *Coloquio* (in the Taylor Collection) are corrected in the *Arte* but almost entirely to values below 1°, while the roll of a ship could make as much difference as 4° to 5° (Taylor, *The Haven-Finding Art*, p. 166).

21. See question 43.

22. This exception is common to all the contemporary Spanish and Portuguese texts.

23. The Portuguese language has preserved this nomenclature.

24. These hours are counted according to the canonical hours of the day, starting at 6:00 A.M. with Prime. None is a "little" hour, 3:00 P.M. by the clock.

25. See Kuhn, *The Copernican Revolution*, pp. 60-63. His entire chapter 2 deals with the "Problem of Planets." The equant was the last of the three devices developed in ancient times to reconcile observed motion of the stars with respect to the sphere. These calculations were of interest mainly to astronomers; Medina's reference is limited to introducing the terms and such facts as he needs to explain eclipses in the next question. It is of interest in this context that José

M. Millás Vallicrosa in *Nuevos estudios sobre historia de la ciencia española*, p. 329, says that Martín Cortés, contemporary of Medina and author of an *Arte de navegar* (see chapter 3 above) cited an Arabic source for the apparent acceleration and deceleration of the planets, the *Liber motu octave sphere* of Tabit ibn Qurra (*Al Andalus* 10 [1945]: 89-108).

26. This reasoning was later adopted by Jerónimo de Chaves. Other writers, like Falero and Cortés, speculated on the abode of condemned souls, who were thought to be the cause of thunder and lightning.

27. Medina's windrose is one of many designed in his time and has been printed in full color for its artistic value in Julio F. Guillén, *Cartographía marítima española*.

28. This description is of great circle sailing.

29. The expression of these differences per degree in leagues is the difficult question which Medina avoided in his *Coloquio* and which a friend added to that text (see chap. 3 above). The numbers are given as a table in Taylor, *The Haven-Finding Art*, p. 164. See also Waters, *Art of Navigation in England*, p. 73.

30. The theories of Anaxagoras and Aristotle are discussed in Gershenson and Greenberg, "How Old Is Science? Two Conflicting Theories of Summer Hailstorms Found in the Science of Antiquity Demonstrate the Antiquity of Science." See also by the same authors, *Anaxagoras and the Birth of Scientific Method* (New York: Blaisdell, 1964), p. 51 and passim.

31. This is the reason for locating hell in the center of the earth as Medina did in the *Coloquio* and other works. The fallen angels had been sent down never to rise again, and the center of the earth is the one place from which every direction is up. The location of the earthly paradise, not discussed in the *Libro*, is mentioned in the *Coloquio* and described as unknown.

32. "Climate," as used by the Greeks, probably referred to the supposed slope of the earth toward the pole, or to the inclination of the earth's axis. It was an astronomical or mathematical term not associated with any idea of physical climate.

33. José Gavira Martín in *La ciencia geográfica española del siglo XVI*, p. 22, gives the traditional climates: *dia-* for the Greek term to traverse; Diameros from Meroe, Ethiopia; Diasiene from Siena in Ethiopia; Diaalexandros from Alexandria; Diarrodes from Rhodes; Diaroma from Rome; Diaborestes from the river Boristenes in Scythia (Carpathians to the Don); Diarrifeo from the mountains "Rifeo," probably of the Urals. The southern climates had the same names prefaced by *anti-*. For the history of the climates, see Ernst Honigmann, *Die Sieben Klimata . . .* (Heidelberg, 1929).

34. This measure of $17\frac{1}{2}$ leagues per degree had already been abandoned by chart makers in 1538. Medina adopted the currently used $16\frac{1}{2}$ leagues measure in his *Arte de navegar*. The problem is most concisely discussed in: Roger Barlow, *A Brief Summe of Geographie*, ed. E. G. R. Taylor, Hakluyt Society, 2d ser., no. 69 (London, 1932), app. 2, called "The Measure of A Degree"; and in Salvador García Franco, *La legua náutica en la edad media* (Madrid, 1957), pt. 3, pp. 199 ff.

35. This question deals with the problem of setting a course from a chart upon which directions are indicated by winds. In setting a course upon the direction found, it was assumed that this could be made good by maintaining a bearing constant with respect to the meridian throughout the voyage. See question and answer 55. Medina did not treat the problem of the flat chart.

36. This color scheme is the usual one used on charts for the principal, half-, and quarter-winds.

Notes on the Translation

Bibliography

Antonio, Nicolás. *Biblioteca hispana nova*. Madrid, 1788.

Archivo General de Indias (A.G.I.). Seville. Secciones: Contratación, Indiferente, Justicia, Patronato.

Arneson, E. P. "The Early Art of Terrestrial Measurement and Its Practice in Texas." *Southwestern Historical Quarterly* 29 (October 1925): 79–87.

Barrántes Maldonado, Alonso. *Ilustraciones de la Casa de Niebla*. 2 vols. In *Memorial histórico español: Colección de documentos, opúsculos y antigüedades que publica la Real Academia de la Historia*, vols. 9–10. Madrid, 1857.

Barreiro-Meiro, Roberto. "La cartografía en tiempos del Emperador." *Revista General de Marina*, Madrid, 1956.

Beaujouan, Guy. "Science livresque et art nautique au XVᵉ siècle." In *Les aspects internationaux de la découverte océanique*. Colloque international d'histoire maritime, 5th, 1960. Paris, 1966.

———, and Poulle, Emmanuel. "Les origines de la navigation astronomique aux XIVᵉ et XVᵉ siècles." In *Le navire et l'économie maritime. . . .* Colloque international d'histoire maritime, 1956. Paris, 1957.

Bensaude, Joaquim. *Histoire de la science nautique portugaise*. Munich, 1914.

———. *Réimpression de critiques étrangères*. Lisbon, 1921.

Blaeu, Willem Janszoon. *Licht der Zeevaert*. Amsterdam, 1608. *The Light of Navigation*. Amsterdam, 1612. Translation reprinted. Amsterdam: Meridian, 1964.

Carande, Ramón. *Carlos V y sus banqueros: La vida económica de España en una fase de su hegemonía, 1516–1556*. 2d ed. 3 vols. Madrid, 1965–67.

Caro Baroja, Julio. "The City and the Country." In *Mediterranean Countrymen: Essays in the Social Anthropology of the Mediterranean*, edited by Julian Pitt-Rivers. The Hague: Mouton, 1963.

Chaunu, Huguette and Pierre. *Séville et l'Atlantique, 1504–1650.* 7 vols. Paris, 1955–59.

Chaves, Alonso de. *Quatri partitu en cosmographía pratica y por otro nombre llamado espeio de navegantes.* Excerpted in C. Fernández Duro, *De algunas obras desconocidas de cosmografía y de navigación.* Madrid, 1895.

Cipolla, Carlo M. *Guns, Sails, and Empires: Technological Innovation and the Early Phases of European Expansion, 1400–1700.* New York: Pantheon, 1965.

Colección de documentos inéditos para la historia de España. 48 vols. Madrid, 1842–95.

Colección de documentos inéditos relativos al descubrimiento, conquista y organización de las antiguas posesiones españolas de América y Oceanía. 42 vols. Madrid, 1864–84.

Colección de documentos inéditos relativos al descubrimiento, conquista y organización de las antiguas posesiones españolas de ultramar. 25 vols. Madrid, 1885–1932.

Cortés, Martín. *Breve compendio de la sphera y del arte de navegar.* . . . Seville, 1551.

Cortesão, Armando. *Cartografía y cartografos portugueses dos seculos XV e XVI.* 2 vols. Lisbon, 1935.

Craster, H. H. E. *The Western Manuscripts of the Bodleian Library.* New York: Macmillan, 1921.

Dominguez Ortiz, Antonio. *Orto y ocaso de Sevilla.* Seville, 1946.

Duarte Leite, J. *Historia dos descobrimentos.* Lisbon, 1958.

Elliott, John H. *Imperial Spain, 1469–1716.* New York: St. Martins, 1964.

Falero (Faleiro), Francisco. *Tratado del esphera y del arte del marear.* Seville, 1535.

Fernández Duro, Cesáreo. *De algunas obras desconocidas de cosmografía y de navigación, y singularmente de la que escribió Alfonso de Chaves á principios del siglo XVI.* Madrid, 1895.

———. *La armada española.* 9 vols. Madrid, 1895–1903.

———. *Disquisiciones náuticas.* 6 vols. Madrid, 1878–81. Vol. 6. *Arca de Noé.* 1881.

Fontoura da Costa, A. *A marinharia dos descobrimentos.* Lisbon, 1940.

García de Cespedes, A. *Regimiento de navegación.* Madrid, 1606.

García Franco, Salvador. *Historia del arte y ciencia de navegar.* 2 vols. Madrid, 1947.

García Mercadal, J. *Viajes de extranjeros por España y Portugal.* Madrid, 1952.

Gavira Martín, José. "La ciencia geográfica española del siglo XVI, Martín Cortés, Martín Fernández de Enciso, Jerónimo de Chaves, Francisco Falero." *Publicaciones de la Sociedad Geográfica Nacional,* no. 7. Madrid, 1931.

Gershenson, D. E., and Greenberg, D. A. "How Old is Science? Two Conflicting Theories of Summer Hailstorms Found in the Science of Antiquity Demonstrate the Antiquity of Science." *Columbia University Forum* 7 (Spring 1964): 24–27.

Gombrich, E. H. *Art and Illusion: A Study in the Psychology of Pictorial Representation.* 2d ed. New York: Pantheon, 1961.

González Palencia, Angel, ed. *Obras de Pedro de Medina.* Clásicos Españoles, vol. 1. Madrid, 1944.

Guevara, Antonio de. *Arte de marear y de los inventores della.* . . . Valladolid, 1539.

Guillén y Tato, Julio F. *Cartografía marítima española.* Madrid, 1943.

———. *Europa aprendió a navegar en libros españoles.* Contribución del Museo Naval de Madrid a la Exposición del Libro del Mar. Barcelona, 1943.

Hall, Marie (Boas). *The Scientific Renaissance, 1450–1530.* London: Collins, 1962.

Haring, C. H. *Trade and Navigation between Spain and the Indies in the Time of the Hapsburgs.* London, 1918.

Humboldt, Alexander von. *Kosmos.* 4 vols. Stuttgart, 1844.

Konetzke, R. *Süd und Mittelamerika.* Frankfort on Main: Fischer Verlag, 1965.

Kuhn, Thomas S. *The Copernican Revolution: Planetary Astronomy in the Development of Western Thought.* Cambridge: Harvard University Press, 1957.

Lamb, Ursula. "The Cosmographies of Pedro de Medina." In *Homenaje a Rodriguez Moñino: Estudios de erudición que le ofrecen sus amigos o discípulos hispanistas norteamericanes.* Madrid, 1966.

———. "The *Quatri Partitu* of Alonso de Chaves, an Interpretation." *Revista da Universidade de Coimbra* 24 (1969): 3–9.

———. "Science by Litigation: A Cosmographic Feud." *Terrae Incognitae* (Amsterdam) 1 (1969): 40–57.

LaPeyre, Henri. *Une famille de marchands, les Ruíz.* Paris, 1955.

Leonard, Irving. *Books of the Brave.* New York, Gordian Press: 1964.

López Martínez, Celestino. "Hermandad de Santa Maria del Buen Aire de la Universidad de Mareantes de Sevilla." In *Anuario de estudios hispano-americanos.* Seville, 1944.

Lynch, John. *Spain under the Hapsburgs, 1516–1598.* Oxford, 1964.

Lynn, Caro. *A College Professor of the Renaissance: Lucio Marineo Siculo among the Spanish Humanists.* Chicago: University of Chicago Press, 1937.

Martínes Hidalgo Terán, J. M. *Historia y leyenda de la aguja magnética.* Barcelona, 1946.

Medina, José Toribio. *Biblioteca hispano-americana.* 7 vols. Santiago, Chile, 1898–1907.

———. *El veneciano Sebastián Caboto al servicio de España.* 2 vols. Santiago, Chile, 1908.

Medina, Pedro de. *Arte de navegar.* . . . Valladolid: Francisco Fernández de Cordova, 1545.

———. *Coloquio de cosmographía* (1543). MS. H. C. Taylor Collection. On deposit. New Haven. Yale University. Beinecke Rare Book Library.

———. *Crónica de los duques de Medina Sidonia* (1561). In *Colección de documentos inéditos para la historia de España.* Vol. 39, pt. 1. Madrid, 1861.

———. *Diálogo de la verdad* (1555). Published as *Libro de la verdad.* In *Obras de Pedro de Medina.* Edited by A. González Palencia. Clásicos Españoles, vol. 1. Madrid, 1944.

———. *Libro de grandezas y de cosas memorables de España* (1548). In *Obras de Pedro de Medina.* Edited by A. González Palencia. Clásicos Españoles, vol. 1. Madrid, 1944.

———. *Regimiento de navegación.* . . . Seville: Juan Canalla, 1552.

———. *Regimiento de navegación.* . . . Seville: Simón Carpintero, 1563. Reprinted in facsimile edition with transcription. *Regimiento de navegación, compuesto por el maestro Pedro de Medina, 1563.* 2 vols. Madrid: Instituto de España, 1964.

———. *Suma de cosmographía* (1550). MS. Madrid. Biblioteca Nacional. Sección MSS. Res. 215.

———. *Suma de cosmographía* (1561). MS. Seville. Cathedral Chapter. Colombina Library. Facsimile edition with foreword by Rafael Estrada. Seville: Diputación Provincial de Sevilla, 1947.

Mexía, Pedro. *Crónica del Emperador Carlos V escrita por su cronista el magnífico caballero Pedro Mexía.* Edited by Juan de Mata Carriazo. 2 vols. Madrid, 1945.

Michel, Henri. "Astrolabistes, géographes et graveurs belges du XVIe siècle." In *La science au seizième siècle.* Colloque international de Royaumont, 1957. Paris: Hermann, 1960.

Millás Vallicrosa, José M. *Nuevos estudios sobre historia de la ciencia española.* Barcelona, 1960.

Mollat, M., ed. *Le navire et l'économie maritime du Moyen-Age au XVIIIe siècle.* Colloque international d'histoire maritime, 2d, 1957. Paris, 1958.

Morison, Samuel Eliot. *Portuguese Voyages to America in the Fifteenth Century.* Cambridge: Harvard University Press, 1940.

Navarrete, Martín Fernández de. *Biblioteca marítima española.* 2 vols. Madrid, 1851.

————. *Disertación sobre la historia de la náutica.* Madrid, 1846.

Otte, Enrique. *Cedulario de la monarquía española relativo a la isla de Cubagua, 1523-1550.* 2 vols. Caracas: Fundación Eugenio Mendoza, 1961.

Pardo de Figueroa, Rafael. *Crítica* [of the *Regimiento de navegación,* 1563]: *Seguida de una ojeada sobre el Arte de navegar (1545) y la Suma de cosmographía (1561). . . .* Cadiz, 1867.

Parry, J. H. *The Age of Reconnaissance: Discovery, Exploration, and Settlement, 1450-1650.* Cleveland: World, 1963.

Picatoste y Rodriguez, Felipe. *Apuntes para una biblioteca científica española del siglo XVI.* Madrid, 1891.

Pike, Ruth. "Seville in the Sixteenth Century." *Hispanic American Historical Review* 41 (February 1961): 1-31.

Pulido Rubio, José. *El piloto mayor de la Casa de Contratación de Sevilla.* Seville, 1923. Revised and enlarged. Seville, 1950.

Rey Pastor, Julio. *La ciencia y la técnica en el descubrimiento de América.* Buenos Aires: Austral, 1942.

Santa Cruz, Alonso de. *Crónica de los reyes católicos.* Edited by Juan de Mata Carriazo. 2 vols. Seville, 1951.

Sanz, Carlos. *Geografía de Ptolomeo.* Madrid, 1959.

Schaefer, Ernesto. "La Universidad de los Mareantes de Sevilla y su intervención en el viage de las flotas a las Indias." *Archivo Hispalense.* 2d ser., no. 14. Seville, 1946.

Strutt, John W. (Lord Rayleigh). "On the Light from the Sky, Its Polarization and Colour." *Philosophical Magazine,* 4th ser. 41 (February 1871): 107-20.

Taylor, E. G. R. *The Haven-Finding Art.* London: Hollis & Carter, 1956.

————. "Instructions to a Colonial Surveyor." *Mariner's Mirror* 37 (1951): 48-63.

————. "The South-Pointing Needle." *Imago Mundi* 8 (1951): 1-8.

————, ed. *Roger Barlowe, A Brief Summe of Geography.* Hakluyt Series 2. Vol. 69. London, 1922.

————, and Richey, M. W. *The Geometrical Seaman.* London, 1962.

Toro Buiza, Luís. "Notas biográficas de Pedro de Medina." In *Revista de estudios hispánicos,* vol. 2, pp. 31-38. Madrid, 1936.

Usher, A. R. "Spanish Ships and Shipping in the Sixteenth and Seventeenth Centuries." In *Facts and Factors of Economic History,* pp. 189-213. Essays in Honor of F. Gay. Cambridge: Harvard University Press, 1932.

Villa, A. M. *Resumen histórico de la Universidad de Sevilla*. Seville, 1886.

Vigneras, L. A. "The Cartographer Diogo Ribeiro." *Imago Mundi* 16 (1962): 76-83.

Vindel, Francisco. *Pedro de Medina y su Libro de grandezas*. Madrid, 1927. Pamphlet, an edition of 100 copies.

Waters, David W. *The Art of Navigation in England in Elizabethan and Early Stuart Times*. New Haven, Yale University Press, 1958.

Winter, H. "The Pseudo Labrador and the Oblique Meridian." *Imago Mundi* 2 (1937): 61-75.

Woodbridge, Hensley C. "A Tentative Bibliography of Spanish and Catalan Nautical Dictionaries, Glossaries, and Word Lists." *The Mariner's Mirror* (Journal of the Society for Nautical Research) 37 (1951): 63-75.

Wroth, Lawrence C. *The Way of a Ship: An Essay on the Literature of Navigation Science*. Portland, Me.: Southworth-Anthoensen, 1937.

Zoubov, Vassili Pavlovitch. "Vitruve et ses commentateurs du XVIe siècle." In *La science au seizième siècle*. Colloque international de Royaumont, 1957. Paris: Hermann, 1960.

Index

PRINTED IN GREAT BRITAIN
AT THE UNIVERSITY PRESS, OXFORD
BY VIVIAN RIDLER
PRINTER TO THE UNIVERSITY